EARTH SCIENCES IN THE 21ST CENTURY

CONTINENTAL SHELF

GEOGRAPHICAL DISTRIBUTION, BIOTA AND ECOLOGICAL SIGNIFICANCE

EARTH SCIENCES IN THE 21ST CENTURY

Additional books in this series can be found on Nova's website under the Series tab.

Additional e-books in this series can be found on Nova's website under the e-book tab.

EARTH SCIENCES IN THE 21ST CENTURY

CONTINENTAL SHELF

GEOGRAPHICAL DISTRIBUTION, BIOTA AND ECOLOGICAL SIGNIFICANCE

LAWRENCE SCANLON
AND
JOYCE L. RANIERI
EDITORS

nova publishers
New York

For permission to use material from this book please contact us:
Telephone 631-231-7269; Fax 631-231-8175
Web Site: http://www.novapublishers.com

Additional color graphics may be available in the e-book version of this book.

Library of Congress Cataloging-in-Publication Data

ISBN: 978-1-62948-330-6

Published by Nova Science Publishers, Inc. †New York

CONTENTS

Preface		**vii**
Chapter 1	The Sea Bottoms of the Continental Shelf of the Naples Bay (Southern Italy): Inferences of Volcanic and Sedimentary Processes on the Stratigraphic Architecture *Gemma Aiello and Ennio Marsella*	**1**
Chapter 2	Geomorphological Characteristics of the Submerged Topography Along the Egyptian Red Sea Coast *Moawad Badawy*	**123**
Chapter 3	Benthic Boundary Layer (BBL) Macrofaunal Community Structure over Oligotrophic Continental Shelves: A New Field Methodological Approach *Panayota Koulouri*	**163**
Index		**191**

PREFACE

In this book, the authors discuss the geographical distribution, biota and ecological significance of the Continental Shelf. Topics include the sea bottoms of the Continental Shelf of the Naples Bay in Southern Italy and its inferences on the volcanic and sedimentary processes on stratigraphic architecture; the geomorphological characteristics of the submerged topography along the Egyptian Red Sea Coast; and the benthic boundary layer (Bbl) of the macrofaunal communities structure over oligotrophic continental shelves.

Chapter 1 – The continental shelf of the Naples Bay has been intensively studied and analyzed through high resolution seismic reflection profiles coupled to morpho-bathymetric data, including Multibeam bathymetry and Side Scan Sonar imagery and sea bottom samples. Marine geological maps of the Naples Bay have been constructed by using an experimental approach to the geological maps, according to guidelines to Italian Marine Cartography suggested by the ISPRA, formerly the Italian Geological Survey. Some criteria and examples of marine geological mapping relative to the geological map n. 465 "Procida" (Naples Bay) both at the scales 1:10.000 and 1:25.000 are here presented. The cartographic approach is based on the recognition of laterally coeval depositional systems, representing portions of system tracts of the Late Quaternary depositional sequence. The integrated geological interpretation of seismic, bathymetric and Sidescan Sonar data has been tied by sea bottom samples and piston cores. The geological structures and the seismic sequences, both volcanic and sedimentary, have been studied in detail through multichannel and single-channel seismic profiles of different resolution and penetration, often integrated with marine magnetics. Seismic interpretation allowed for the reconstruction of stratigraphic and structural setting of shelf-slope successions, separated by tectonic and eustatic regional unconformities. In the Naples Bay the Dohrn and Magnaghi canyons, eroding the slope up to –1000 m of water depth, represent main morpho-structural lineaments at the boundary between the sedimentary units of the eastern shelf and the volcanic units of the western shelf, in correspondence to the Phlegrean Fields, Ischia and Procida volcanic complexes. The marine geological maps, so realized, show the distribution of several lithostratigraphic units cropping out at the sea bottom and of the main morphological lineaments. The stratigraphic units belong to the Late Quaternary Depositional Sequence, showing the spatial and temporal geological evolution and the lateral and vertical migration of marine coastal, continental shelf and slope depositional environments during the Late Pleistocene-Holocene glacio-eustatic cycle. The stratigraphic succession studied through the geological survey has recorded the variations of the accommodation space of the Late Quaternary deposits during the last 4th order glacio-eustatic cycle, ranging between 128 ky B.P. (Tyrrhenian stage) and the present (isotopic stage 5e). The geomorphology and seismic stratigraphy of the continental shelf and slope appear strongly controlled by the interplay of volcanism and canyon formation acting in correspondence to the Magnaghi and Dohrn axes. Mound-shaped morphological highs genetically related to the emplacement of the Neapolitan Yellow Tuff volcanic unit (13 ky B.P.), widely cropping out onshore, Holocene reworked sediments and sea-bottom creep appear on acoustic Chirp profiles in the inner sectors of the bay. These features appear to be related to the interaction among volcanic, tectonic and sedimentary processes conditioning the geological evolution of the Naples Bay during the Late Quaternary. On the contrary, the sedimentation over the shelf at the southern edge of the bay (Sorrento Peninsula-Capri island) seems less influenced by volcanic activity and seabed features include Late Pleistocene regressive sand bodies and Holocene patch reefs, coastal dunes and depositional terraces.

Chapter 2 – A long time ago the Red Sea was only known by small-scale bathymetric, magnetic anomaly maps and a few seismic reflection or refraction profiles. Therefore, detection of the major submerged coastal features was unattainable. This study is based on the integration of different data sets of topography and bathymetry (e.g. the global bathymetry data set, the SRTM DTED-2, the soviet military topographic maps of scale 1:200.000 and the US army topographic maps of scale 1:250.000) to reveal the main submarine landforms that marked the continental shelf and its related slopes along the Egyptian Red Sea Coast from latitude 27°43`N to the Egyptian-Sudanese border at latitude 22°00`N. The study revealed the major types of coral reefs along the Egyptian Red Sea Coast. It is deduced that the continental shelf is noticeably influenced by the surface fault system extending eastward into the main Red Sea depression, showing the continental edge mostly like a fault-scarp of ~60° anticlockwise fault plane. Sea ridges and subbasins were distinguished at the lower toe of the continental slope, which seem to be a result of a regional fold system. Two sea peaks of extinct volcanoes were recognized. Two types of submarine canyons were recognized as deep incised Messinian canyons and shallow canyons. The deep incised canyons (~500 m bsl) carve the continental edge with remarkable steep walls. They might be formed as a result of the Messinian event (~5.59 Ma). The shallow canyons are mostly developed during the Pleistocene lower sea level (~90–130 m bsl) where the major wadis cut their water courses through the continental shelf. Some individual submerged deltas were identified, showing a close relationship with the present-day drainage system, although they were supposed to be produced by an ancestor drainage system. Notable submarine terraces were recognized at depths 20–25, 50–75, and 100–120 m bsl that are in agreement with the generalized global curve of sea-level rise since the LGM (~23–18 ka bp). It is expected that results of this study will be helpful for shedding light on the geomorphologic evolution of the continental shelf and its surrounding landmass.

Chapter 3 – The marine sediment-water interface is one of the most extensive habitats on the planet occupying 75% of the Earth's surface. Fluxes of materials across the sediment-water interface and the mechanisms that mediate and constrain those fluxes are likely to have global importance. Because of differing ecosystem structure above and below the sediment-water interface, the ecologists who study these domains use different techniques and ask different research questions leading often to scientific isolation. Consequently, the significance of the fauna related to the sediment-water interface is underestimated, especially in studies of benthic-pelagic coupling related to energy fluxes and also in holistic approaches of the marine ecosystem. The portion of the sediment and the water column that is influenced by the sediment-water interface is characterized by strong gradients of flow as well as dissolved and particulate matter. This dynamic zone, also known as benthic boundary layer (BBL), can be extended from a few centimetres to tens of metres, depending both temporally and spatially on various physical and biogeochemical processes whereby organisms on either side of the sediment-water interface interact. BBL supports both benthic and pelagic, mostly macrofaunal, organisms classified in different groups according to their different mobility and bottom dependence. BBL species (mainly crustaceans) are considered to be key taxa in marine food webs as most of them exploit a large diversity of food resources, mobilizing organic and detritus particles and associated microbes or preying on meiobenthos and plankton, while at the same time constituting an important food source for demersal fish, epibenthic crustaceans and cephalopods, of high commercial value. For the study of BBL macrofauna, specially designed samplers, generally known as hyperbenthic or suprabenthic sledges, have been used over the last 40 years. Nevertheless, there are still *practical* technical difficulties in sampling efficiently the lowest layer of the water column above the seabed where most of these organisms are concentrated. In the framework of an EU-funded project, a new sampling device for the study of BBL macrofaunal communities was developed and applied experimentally in the field. This novel sampling technique included, as a complementary application, a specially designed facility attached in the frontal part of a modified hyperbenthic sledge for the artificial resuspension of the surface sediment. This facility provided additional quantitative information on the BBL macrofaunal populations concentrated in the lowest layer of the water column, just above the seabed surface, allowing a more comprehensive view of the role of this particular macrofauna in the marine food web to be built. The application of this novel sampling approach gave new insights towards a better understanding of the role of biodiversity related to the sediment-water interface with respect to nutrient regeneration, carbon cycling and energy transfer to higher trophic levels as well as to relevant bottom trawling effects.

In: Continental Shelf
Editors: L. Scanlon and J. L. Ranieri
ISBN: 978-1-62948-330-6

Chapter 1

The Sea Bottoms of the Continental Shelf of the Naples Bay (Southern Italy): Inferences of Volcanic and Sedimentary Processes on the Stratigraphic Architecture

Gemma Aiello* and Ennio Marsella
Istituto per l'Ambiente Marino Costiero (IAMC),
Consiglio Nazionale delle Ricerche (CNR), Calata Porta di Massa,
Porto di Napoli, Napoli, Italy

Abstract

The continental shelf of the Naples Bay has been intensively studied and analyzed through high resolution seismic reflection profiles coupled to morpho-bathymetric data, including Multibeam bathymetry and Side Scan Sonar imagery and sea bottom samples. Marine geological maps of the Naples Bay have been constructed by using an experimental approach to the geological maps, according to guidelines to Italian Marine Cartography suggested by the ISPRA, formerly the Italian Geological

* Corresponding author: Dott. Gemma Aiello, Full-time researcher; Istituto per l'Ambiente Marino Costiero, Consiglio Nazionale delle Ricerche (CNR), Calata Porta di Massa, Porto di Napoli, 80133 Napoli; Tel. 39-81-5423820; Fax: 39-81-5423888; Email: gemma.aiello@iamc.cnr.it.

Survey. Some criteria and examples of marine geological mapping relative to the geological map n. 465 "Procida" (Naples Bay) both at the scales 1:10.000 and 1:25.000 are here presented. The cartographic approach is based on the recognition of laterally coeval depositional systems, representing portions of system tracts of the Late Quaternary depositional sequence. The integrated geological interpretation of seismic, bathymetric and Sidescan Sonar data has been tied by sea bottom samples and piston cores. The geological structures and the seismic sequences, both volcanic and sedimentary, have been studied in detail through multichannel and single-channel seismic profiles of different resolution and penetration, often integrated with marine magnetics. Seismic interpretation allowed for the reconstruction of stratigraphic and structural setting of shelf-slope successions, separated by tectonic and eustatic regional unconformities. In the Naples Bay the Dohrn and Magnaghi canyons, eroding the slope up to – 1000 m of water depth, represent main morpho-structural lineaments at the boundary between the sedimentary units of the eastern shelf and the volcanic units of the western shelf, in correspondence to the Phlegrean Fields, Ischia and Procida volcanic complexes. The marine geological maps, so realized, show the distribution of several lithostratigraphic units cropping out at the sea bottom and of the main morphological lineaments. The stratigraphic units belong to the Late Quaternary Depositional Sequence, showing the spatial and temporal geological evolution and the lateral and vertical migration of marine coastal, continental shelf and slope depositional environments during the Late Pleistocene-Holocene glacio-eustatic cycle. The stratigraphic succession studied through the geological survey has recorded the variations of the accommodation space of the Late Quaternary deposits during the last 4th order glacio-eustatic cycle, ranging between 128 ky B.P. (Tyrrhenian stage) and the present (isotopic stage 5e). The geomorphology and seismic stratigraphy of the continental shelf and slope appear strongly controlled by the interplay of volcanism and canyon formation acting in correspondence to the Magnaghi and Dohrn axes. Mound-shaped morphological highs genetically related to the emplacement of the Neapolitan Yellow Tuff volcanic unit (13 ky B.P.), widely cropping out onshore, Holocene reworked sediments and sea-bottom creep appear on acoustic Chirp profiles in the inner sectors of the bay. These features appear to be related to the interaction among volcanic, tectonic and sedimentary processes conditioning the geological evolution of the Naples Bay during the Late Quaternary. On the contrary, the sedimentation over the shelf at the southern edge of the bay (Sorrento Peninsula-Capri island) seems less influenced by volcanic activity and seabed features include Late Pleistocene regressive sand bodies and Holocene patch reefs, coastal dunes and depositional terraces.

1. INTRODUCTION

The inferences of volcanic and sedimentary processes on the stratigraphic architecture of the Naples Bay are here shown and discussed based on the techniques of marine geology and sequence stratigraphy. The sea bottoms of the continental shelf of the Naples Bay have been studied in detail through a marine geological survey of the Campania region (CARG Project) for the construction of experimental geological maps committed from the National Geological Survey of Italy, now ISPRA (Istituto Superiore Protezione Ricerca Ambientale). Later, from the 2003, the Region Campania, Sector of Soil Defense, committed to the sa me Institute also the marine geological mapping of the Naples and Salerno bays at the scale 1:10.000.

Several geological and geophysical surveys of the continental shelf and slope of the Naples Bay have been carried out. In particular, a high resolution Multibeam bathymetry of the Naples Bay allowed for the construction of a marine DEM (Digital Elevation Model) of the area, giving a detailed image of the morpho-structures at the sea bottom (D'Argenio et al., 2004; Figure 1). Moreover, Sidescan Sonar acoustic profiles covering the whole Naples Bay have been acquired for the construction of photomosaics of the sea bottom. The Sidescan Sonar photomosaics and the Multibeam bathymetry represented the base for the marine geological mapping.

The integrated geological interpretation of seismic, bathymetric and Sidescan Sonar data has been tied by sea bottom samples and piston cores. The geological structures and the seismic sequences, both volcanic and sedimentary in nature, which characterize the Naples Bay at a regional scale, have been the object of detailed studies carried out by using multichannel and single channel seismics of various resolution and penetration, often integrated with marine magnetics (Aiello et al., 2001; Secomandi et al., 2003; Aiello et al., 2004; 2005; 2009a; 2009b; 2011a; 2011b; 2012a; 2012b; Ruggieri et al., 2007).

The interpretation of high resolution seismic reflection profiles (Subbottom Chirp, Sparker and Watergun) has supported the reconstruction of stratigraphic and structural framework of the continental shelf and slope successions. The seismo-stratigraphic analysis allowed to distinguish the main volcanic and sedimentary seismic units, separated by regional unconformities, tectonically and eustatically controlled. The Dohrn and Magnaghi canyons, eroding the slope up to 1000 m of water depth, represent in the Naples Bay important morpho-structural lineaments, at the boundary between the sedimentary units of the eastern shelf of the Naples Bay and the volcanic units of the western shelf, in correspondence to the Ischia and Procida islands.

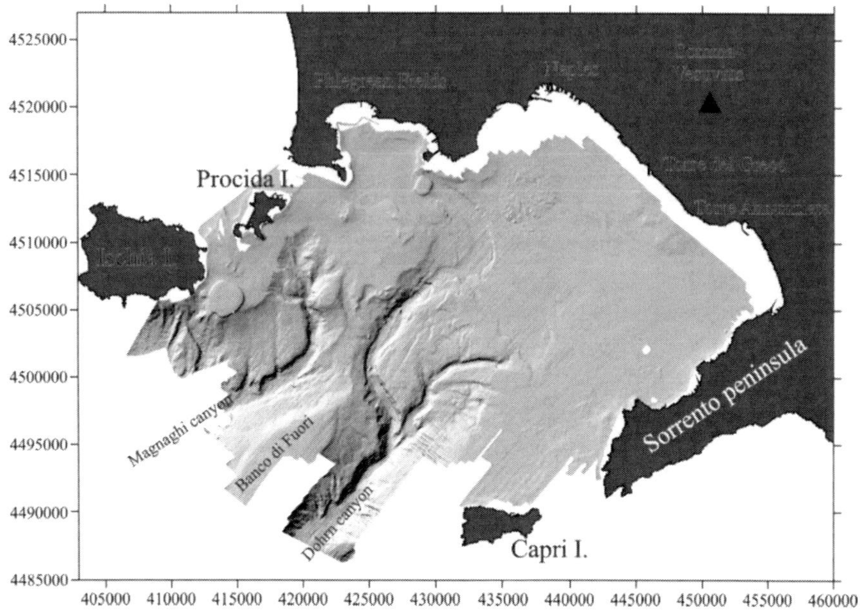

Figure 1. Sketch map showing the location of the study area and the main morphological lineaments in the Gulf of Naples reported from Multibeam bathymetry (modified after Aiello et al., 2005).

The complex stratigraphic architecture of the Naples Bay has revealed, during the Late Quaternary, a strong geological control of volcano-tectonic processes in triggering submarine gravity instabilities. This regional geological framework didn't allow a simple application of principles and techniques of seismic and sequence stratigraphy, as described in the guidelines for the redaction of marine geologic cartography (Catalano et al., 1996; Fabbri et al., 2002). The realized cartographic approach, however experimental, is based on the recognition of laterally coeval depositional systems, representing portions of system tracts of the Late Quaternary Depositional Sequence (Catalano et al., 1996).

The seismic units have been later interpreted in terms of depositional sequences and corresponding unconformities have been interpreted as Type 1 or Type 2 sequence boundaries and/or as local unconformities, mainly at the top of relic volcanic edifices or at the top of volcanic seismic units (D'Argenio et al., 2004; Aiello et al., 2005; Ruggieri et al., 2007).

The marine geological map, realized based on the previously mentioned criteria shows the distribution of several lithostratigraphic units cropping out at the sea bottom and of the main morphological lineaments, based on the CARG guidelines for the realization of marine cartography (Catalano et al., 1996; Fabbri et al., 2002).

The main stratigraphic units individuated through the analysis of the sea bottom sediments belong to the Late Quaternary Depositional Sequence; in this sequence it is possible to recognize the space and time evolution and the lateral and vertical migration of marine coastal, continental shelf and slope depositional environments in the Late Pleistocene-Holocene glacio-eustatic cycle. The stratigraphic succession studied through the geological survey has recorded the variations of the accommodation space of the Late Quaternary deposits during the last 4th order glacio-eustatic cycle, ranging between 128 ky B.P. (Tyrrhenian stage) and the present (isotopic stage 5e; Catalano et al., 1996).

One aim of the cartography has been the cartographic representation of the lithofacies associations, whose grouping form the "depositional elements" (which are portions of system tracts), in relation to the morpho-structural lineaments recognized through the geological interpretation of the geophysical data and the dynamics of depositional environments.

In this way, we tried to realize an integration between classical stratigraphic approach, sequence stratigraphic approach and characterization of actual and recent depositional elements. The volcanic activity, which has mainly controlled the stratigraphic architecture of the Naples Bay prevented a classical stratigraphic approach in the marine geological mapping, which has been realized taking into account both the associations of depositional systems and the interstratified volcanic bodies (volcanites and volcaniclastites).

Moreover, this cartographic approach allowed to obtain informations comparable with other sectors of the Italian continental margins. The last Quaternary sea level rise, having an excursion of about 120 m and a maximum rate in the order of 10 m/1000 years has left a stratigraphic signature on the morphological and stratigraphic framework of most continental margins of the world (Chappell and Shackleton, 1986). The deposits associated to this process are strongly different in the various areas, as a function of different sedimentary supply, morphological framework and oceanographic regimes; to map these deposits allows to correlate the unconformities (erosional and-or depositional) and to compare the facies, the internal geometries and the thickness of the deposits recording the sea level rise in a differential way on the several margins.

2. Geologic Setting

The structural framework of sedimentary basins along the Tyrrhenian margin of Campania and their regional subsidence have been intensively studied based on both field geology and on offshore and onshore seismic and well data (D'Argenio et al., 1973; Finetti and Morelli, 1973; Ippolito et al., 1973; Ortolani and Aprile, 1978; Fabbri et al., 1981; Bartole et al., 1983; Bartole, 1984; Trincardi and Zitellini, 1987; Mariani and Prato, 1988). Nonetheless this, the lack of deep geothermal wells in the Gulf of Naples makes more complex the reconstruction of the age of basin formation and the chronological calibration of seismic sequences filling the basin.

The Gulf of Naples represents an excellent natural laboratory, in which basin filling recorded the interplay of tectonic, volcanic and depositional/ erosional processes, together with the tectonic uplift and deformation of adjacent onshore areas during the Pleistocene. The stratigraphic architecture of continental shelf and upper slope settings appears strongly influenced by the volcaniclastic input of centers of the Phlegrean Fields, Somma-Vesuvius and Ischia and Procida islands (Rosi and Sbrana, 1987; Santacroce, 1987; Vezzoli, 1988). Rapid lateral variations between acoustically-transparent seismic units, interpreted as volcanic deposits, which erupted during multiphase volcanic activity of the Phlegrean Fields, Ischia and Procida islands complexes and regularly stratified sedimentary units supplied by the palaeo-Sarno river mouth and by the Sorrento Peninsula tectonic uplift occur.

The continental shelf of the Gulf of Naples has a variable width, ranging between 2.5 km (offshore the western sector of the Capri island) and 10-15 km (offshore the Sorrento coast: Figure 1). Such a submarine topography is controlled by the interactions between subaerial and submarine volcanism, strongly involving the gulf during the Late Pleistocene and the linear erosion and sediment drainage along the main axes of the Dohrn and Magnaghi canyons.

The eruption centers occurring on the islands of Procida, Vivara and Ischia have an age ranging between 150 kyr and historical times (Rosi and Sbrana, 1987; Vezzoli, 1988). The volcanic activity in these islands played a major role in the formation and the activity of the Dohrn canyon western branch and of the Magnaghi canyon, since they drained a great volcaniclastic input on the slope during major eruptive phases. Multi-phase submerged volcanic edifices around Ischia island (i.e. Ischia Bank) and monogenic edifices in the Phlegrean Fields offshore (Pentapalummo Bank, Nisida Bank, Miseno Bank; Latmiral et al., 1971; Pescatore et al., 1984; Fusi et al., 1991;

Figure 1) are also known. These banks represent volcanic relic morphologies characterized by polycyclic erosional surfaces cropping out at the sea bottom, eroding volcanic deposits and covered by Holocene sediments (Pentapalummo Bank, Ischia Bank) or by a thick Holocene sedimentation.

In the center of the Gulf of Naples, between Capri and Ischia islands, a NE-SW trending morpho-structural high (Di Fuori Bank or Bocca Grande Bank) occurs. This bank looks like an asymmetrical ridge with the southeastern flank steeper than the north-western one. It results from the regional uplift and tilting of Mesozoic carbonatic acoustic basement along the Sorrento Peninsula – Capri island structural high (Figure 1).

The volcanic areas of the Phlegrean Fields and of the Ischia, Procida and Vivara islands share the same tectonic setting and a common origin of their alkali-potassic magmas, which characterize the Campania volcanoes. These magmas are subdivided into an ultra-potassic series (leucitites and leucitic phonolites, in the Somma-Vesuvius volcanic complex) and a potassic series (trachybasalts, latites, trachytes and phonolites, in the Ischia and Procida islands and in the Phlegrean Fields). The whole region pertains to the eastern Tyrrhenian margin, located on a thinned continental lithosphere, transitional with respect to the oceanic sectors of the Tyrrhenian bathyal plain (Kastens et al., 1986; 1988).

The lithospheric extension, accompanying the opening of the Tyrrhenian basin from Pliocene to recent times, has controlled a strong kinematic and translational tectonics in the upper crust, where the uprising of highly differentiated magmas occurred, mostly along Apenninic (NW-SE) and counter-Apenninic (NE-SW) structural lineaments. This evidence among others is testified by the drilling of andesitic rocks in the subsurface northwards of the Phlegrean area (Ortolani and Aprile, 1978; Aiello et al., 2011a).

The Ischia island extends over a surface of 42 kmq and reaches a maximum height of about 787 m at the Epomeo Mt., resulting from the volcano-tectonic uplift of the corresponding caldera during the last 30 ky (Orsi et al., 1991; Acocella et al., 1997; Acocella and Funiciello, 1999). On the Ischia island, volcanic deposits resulting from both explosive and effusive eruptions extensively crop out, building volcanic edifices. Some of them are still well preserved, while others are completely eroded or buried (Forcella et al., 1983; Gillot et al., 1982; Luongo et al., 1987; Vezzoli, 1988). Also landslide deposits, derived from the accumulation and cementation of fragments of pre-existing volcanic rocks, are widely documented (Guadagno and Mele, 1995; Mele and Del Prete, 1998; De Vita et al., 2006; 2007; Iovino and Perriello Zampelli, 2007; Vingiani and Terribile, 2007; de Alteriis and Violante, 2009).

The main volcanologic event is the eruption of the Green Tuffs of the Epomeo Mt., that occurred about 55 ky ago, controlling the downthrowing of the central sector of the island due to the formation of a caldera (Orsi et al., 1991). The volcanic activity of the island has been influenced by complex geological processes of calderic resurgence that started about 30 ky ago. The uplift, related to this last event, has been evaluated in the order of 800-1100 m (Barra et al., 1992a; 1992b).

Three main eruptive cycles have been distinguished based on radiometric datations (Civetta et al., 1991) considering the trend of both Sr isotopic composition and trace elements. Each eruptive period was characterized by magmatic fractioning, allowing for compositional variations of the volcanic products.

The first period (55-33 ky B.P.) started with the eruption of the Epomeo Green Tuffs, characterized by ignimbrites, which deposited both offshore (in an area actually located in the central sector of the island) and now exposed onshore at the Epomeo. The subaerial volcanic products of the Epomeo Green Tuffs crop out at Monte Vico, Sant'Angelo and along the Scarrupata di Barano. After their outbreak, the volcanic activity continued with several explosive eruptions, lasting up to 33 ky.

The corresponding volcanites crop out both along the coastal cliffs of Sant'Angelo and Punta Imperatore (southwestern Ischia) and at Citara and Monte Vico (Citara-Serrara Fontana Formation). Some authors have differentiated these volcanic products into several lithostratigraphic units, erupted by different volcanic centers located in the southwestern and northwestern sectors of Ischia (Chiesa et al., 1985). The second period (28-18 ky B.P.; Civetta et al., 1991) started with the eruption of the Grotta di Terra volcanic center (28 ky B.P.), whose products, fed by trachybasaltic magmas, are exposed along the southeastern coasts of the island. The volcanic activity continued up to 18 ky B.P. the corresponding volcanites crop out at Grotta del Mavone, Monte di Vezzi, Sant'Anna, Carta Romana and between Punta Imperatore and Sant'Angelo (Vezzoli, 1988).

The third period began at about 10 ky B.P. (Civetta et al., 1991) after a relatively long stand of volcanic activity and continued up to historical times, with the so-called Arso lava flow (1302 D.C., Vezzoli, 1988). This period was characterized also by lava domes and by explosive activity, generating tuff rings and pyroclastic fall deposits.

During this period, many eruptive centers were active in the Ischia graben, a depression located in the central sector of Ischia (Orsi et al., 1994). The Zaro lava flow and the pyroclastic deposits cropping out at Punta del Chiarito (Vezzoli, 1988) pertains to this period of activity.

The last eruption in the Ischia island occurred in February of 1302, when a crater opened in the Fiaiano area and large lava flow erupted (Arso lava flow), that reached the beach between the Ischia harbour and Ischia Ponte and destroyed the ancient town of Geronda. At present, only a fumarolic and hydrothermal activity has been recorded.

The tectonics of the island is characterized by NW-SE and NE-SW trending extensional faults., Plio-Quaternary in age (Acocella and Funiciello, 1999; Acocella et al., 2004). Field data have been collected to explain both the tectonic setting of Ischia island, interpreted as a resurgent dome uplifted for almost 800 m in the last 33 ky and the deformational patterns related to this resurgence. NW-SE and NE-SW trending extensional fault systems predominate all over the island, suggesting a close relationship with the extensional structures of southern Apennines (D'Argenio et al., 1973; Bigi et al., 1992).

Marine exploration of the Ischia offshore showed a great improvement in the last 10 years, due to surveys carried out by the IAMC-CNR of Naples in the frame of the GNV (National Group of Volcanology) activity and CARG projects. These surveys resulted in a Multibeam coverage that allowed the exploration of the whole area surrounding the island (Chiocci and de Alteriis, 2006; Aiello et al., 2009a; 2009b; 2011b; 2012a; 2012b). The Ischia offshore is characterized by several monogenic edifices, aligned along the NW-SE system of faults that connects the southeastern sector of Ischia to the island of Procida and to the Phlegrean Fields (de Alteriis and Toscano, 2003; de Alteriis et al., 2006). These edifices have been mainly formed through subaqueous explosive eruptions, documented by the hialoclastites from a potassic magma of latitebasaltic-latitic composition (Di Girolamo and Rolandi, 1976; Di Girolamo et al., 1984). Marine geophysical surveys revealed collapsed volcanic edifices located to the W and SW of Ischia island, a sector of seafloor characterized by strong residual magnetic anomaly fields. A westward submerged continuation of the Ischia volcanic structure, representing a remnant of the emerged top of a larger, E-W trending volcanic complex, has been also inferred, based on marine geophysical data (Bruno et al., 2002).

The Phlegrean Fields are a volcanic district surrounding the western part of the Gulf of Naples, where volcanism has been active since at least 50 ky (Rosi and Sbrana, 1987). They correspond to a resurgent caldera (Rosi and

Sbrana, 1987; Di Vito et al., 1999; Nunziata et al., 1999; Orsi et al., 2002; 2009; D'Argenio et al., 2004; Deino et al., 2004; Milia and Torrente, 2007; Blockley et al., 2008), with a diameter of 12 km (Phlegrean caldera) and resulting from the volcano-tectonic collapse induced from the eruption of the pyroclastic flow deposits of the Campanian Ignimbrite (35 ky B.P.). Coastal sediments ranging in age from 10 to 5.3 ky crop out at a 50 m altitude on the sea level in the marine terrace of "La Starza" (Gulf of Pozzuoli), indicating a volcano-tectonic uplift of the caldera center (Dvorak and Mastrolorenzo, 1991; Morhange et al., 2005).

Monogenic volcanic edifices representing the offshore rim of the caldera center (Pentapalummo Bank; Miseno Bank; Nisida Bank) are well known from a geological and volcanological point of view (Aiello et al., 2001; 2005; Milia, 2010). These banks represent relict volcanic morphologies characterized by polycyclic erosional surfaces cropping out at the sea floor, eroding volcanic deposits and covered by Holocene sediments, highly varying in thickness.

The geological setting of the Phlegrean Fields and their stratigraphy have been summarized by Rosi and Sbrana (1987). The Quaternary volcanic area of the Phlegrean Fields is located in a central position within the graben of the Campania Plain (D'Argenio et al., 1973; Bigi et al., 1992). The main structural element is represented by a wide caldera (the Phlegrean caldera – Figure 3), individuated after the volcano-tectonic collapse following the emplacement of the Campanian Ignimbrite, a large pyroclastic flow, which covered the whole plain about 35 ka ago (Barberi et al., 1978; Rosi et al., 1996; 1999; Civetta et al., 1997; Pappalardo et al., 1999; Signorelli et al., 2001; Fedele et al., 2002; 2008; Rolandi et al., 2003; Marianelli et al., 2006; Pyle et al., 2006). Within the Phlegrean caldera and along its margins, the volcanic activity continued into historical times.

The pre-calderic volcanic activity developed in correspondence with small and scattered volcanic centers, erupting trachytic pyroclastic and lavas. The corresponding outcrop located at M.te di Procida coastal cliff exhibits a thick subaerial pyroclastic sequence with several interbedded paleosols. It includes four main pumice fall beds, a welded pyroclastic flow deposit, and dark lapilli and ashes erupted by the Fiumicello volcanic center (Procida island).

The post-calderic volcanic activity developed during four main phases, individuated through radiometric age determinations and paleogeographic evolution, in a time interval spanning from 35 ky to historical times. The oldest one (35-11.5 ky) has been established based on deep well stratigraphy in the Phlegrean area (Mofete and SanVito geothermal area; Aiello et al., 2012c) and includes thick submarine volcaniclastic deposits (tuffites) interlayered with minor lava bodies, covered by large outcrops of yellow tuffs, reaching their maximum thickness in the city of Naples (Neapolitan Yellow Tuff; Scarpati et al., 1993). Stratified yellow tuffs form the proximal facies of typical tuff-cones cropping out in the surroundings of Naples and Pozzuoli towns. The Neapolitan Yellow Tuffs crop out mainly in the peripheral areas of the Phlegrean caldera and have been erupted by several volcanic centers. A number of vents have been identified in field outcrops of the town of Naples. The second, third and fourth eruptive phases spanned from 10.5 ky to A.D. 1538 and were mainly subaerial. The corresponding volcanic deposits are separated by well stratified paleosols, emplaced within the caldera and along its outer margins (Deino et al., 2004; Pabst et al., 2008).

Among the main volcano-tectonic structures of the Phlegrean Fields there is La Starza terrace (Rosi and Sbrana, 1987), a marine terrace near the centre of the caldera, characterized by littoral deposits overlain by thin subaerial pyroclastic deposits. It is articulated into two levels separated by a step; the upper one, on which the town of Pozzuoli is superimposed, reaches a height of 50-54 m, while the lower one developed at 40 m.

A sketch geomorphological map of the Pozzuoli area has been constructed to improve the geological knowledge of the onshore area (Figure 2). The main morphological lineaments are the lava domes, the crater rims, the collapse rims, the Late Holocene sea cliff, the middle Holocene sea cliff, the faults older than the marine ingression, the faults affecting the La Starza terrace, the rims of subaerial terraces and the local elevation of terraces. The Holocene marine terrace of La Starza is bounded by an abandoned sea cliff (Cinque et al., 1985; 1997). The succession exposed in this cliff overlies a volcanic substratum of yellow tuffs, including fossiliferous littoral deposits alternating with subaerial pyroclastic deposits and paleosols (Cinque et al., 1985; 1997). The lowest marine interval exposed in the cliff has been radiometrically dated at 10.5 ka (Insinga et al., 2002); an overlying paleosol gave an age of about 8 ka and the uppermost beach deposits an age of 4.6 ka. The continental intervals of the succession were deposited during periods of uplift having rates greater than sea-level rise. The final uplift of the terrace (about 30 m in a few centuries) was probably related to the caldera resurgence, accompanying the onset of a new period of strong volcanic activity of the Phlegrean Fields (Cinque et al., 1997; Insinga et al., 2002).

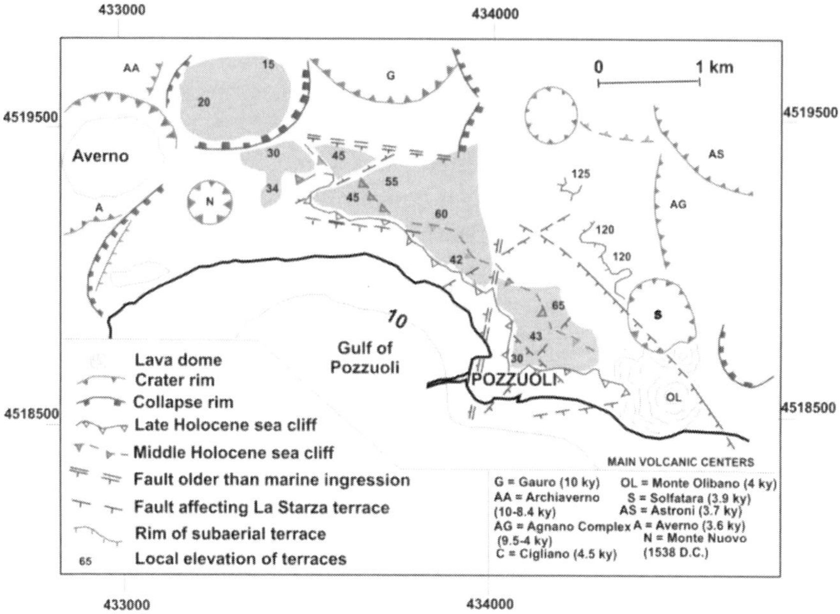

Figure 2. Sketch geomorphic map of the Pozzuoli area (Phlegrean Fields). Main volcanic centers have been reported (modified after Aiello et al., 2012).

The stratigraphic study of five deep cores drilled in the Gulf of Pozzuoli along with the radiometric age dating (^{14}C AMS) of two peat layers and a number of geotechnical tests allowed for the reconstruction of the stratigraphic framework and the determination of physical and mechanical properties of the Holocene succession beneath the sea floor (Insinga et al., 2002).

The Holocene succession is characterized by transgressive volcanic sandy deposits, including a thick pyroclastic unit. Samples collected from the peat layers yelded radiometric ages of 3560 and 7815 cal year B.P. (Insinga et al., 2002). This suggested a correlation between the cored peat layers and two paleosols documented on land that are associated with periods of quiescence during the volcanic activity of the Phlegrean Fields. The Holocene deposits can be correlated with distinct seismo-stratigraphic units identified in the subsurface of the southern and central sectors of Pozzuoli Bay. They can be also correlated with the section cropping out onshore, along the erosional slope of La Starza terrace, where the marine Holocene sequence has been uplifted about 40 m above sea level in the last few thousand years (Insinga et al., 2002).

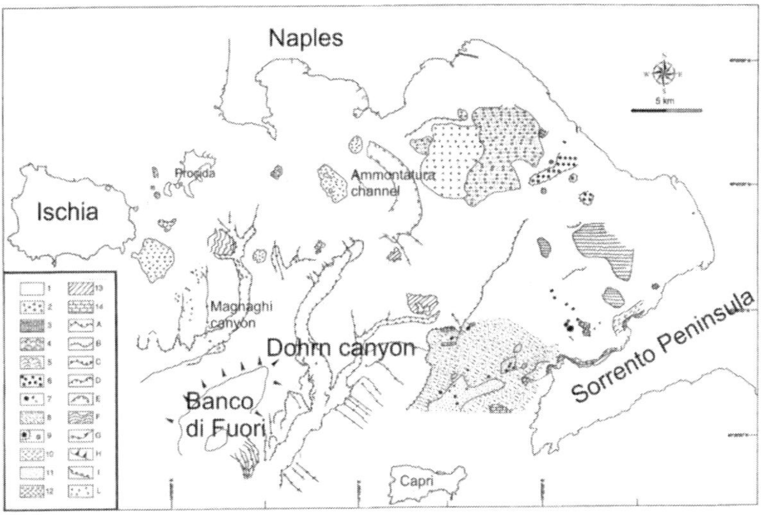

Figure 3. Geomorphological map of the Bay of Naples (modified after Aiello et al., 2001). Deposits: Late Pleistocene-Holocene: 1 post-glacial drape; 2 reworked sediments (mud flows, lahars); 3: sediments deformed by creep; 4 sand lobes; 5 coastal dunes (buried by post-glacial drape); 6 mounded-shaped pyroclastic morphological highs; 7 algal patch reefs (post-glacial); 8 coastal regressive sands (buried by post-glacial drape); 9 submerged or buried parasitic vents or maars (Somma-Vesuvius); 10 volcanic edifices (tuff cones, domes) and their remnants (Middle-Late Pleistocene); 11 coastal regressive sands (last glacial); 12 volcanic edifices (tuff cones, domes) and their remnants (Middle-Late Pleistocene); 13 remnants of the Middle-Late Pleistocene continental shelf (Cretaceous); 14 outcrops of carbonatic acoustic basement (Meso-Cenozoic). Landforms. A: shelf break; B: terrace rim; C: canyon wall; D: slides and scars along canyon walls; E: slump scars; F: gravity slumps; G: drainage patterns; H: Banco di Fuori slope; I: Ammontatura channel rim; L: debris flow outrunner blocks.

3. Data and Methods

Most of the data here presented have been acquired in the frame of three oceanographic cruises organized by the Institute for the Marine Coastal Environment of the National Research Council, Naples, Italy during the 1996-1998 time span. The first cruise (R/V Urania, GMS97-1; October-November, 1996) was dedicated to the acquisition of a high-resolution Multibeam survey of the Bay of Naples (below the – 50 m isobath) and of a narrow belt south of the Sorrento Peninsula. An ELAC Nautik Bottomchart MK2 Multibeam system was adopted with a bathymetric range of – 50/3000 m. The ELAC system is equipped with a transducer of 256 beams arranged to generate a 15 to 150° fan width; the vertical accuracy is about 0.25% of water depth whereas the spatial resolution (the size of the footprint) ranges from 1 to 5 m. The maximum horizontal coverage is 7.5 times the water depth. During the acquisition CTD profiles have been collected every 6 hours. The line spacing has been set to obtain a variable overlap (from 30% up to 100%) between swaths. During the acquisition about 1500 km of Subbottom Chirp profiles have been recorded. During the second cruise (GMS98-1 – May 1998), 110 cores (for a total of more than 450 m), 18 dredges and 41 box-cores have been collected over the area coupled to a grid of more than 500 km of single-channel Sparker 1kJ and 4kJ and magnetometric profiles.

4. Morphobathymetric Results

4.1. General Setting

A geomorphological map of the Bay of Naples has already been constructed (Aiello et al., 2001) and is shown in Figure 3. It is based on Multibeam and Subbottom Chirp surveys collected in the frame of research programmes of marine geological mapping. During the Late Pleistocene-Holocene the post-glacial drape, the reworked sediments (mud flows, lahars), the sediments deformed by creep, the sand lobes, the coastal dunes (buried by post-glacial drape), the pyroclastic lumps genetically related to the upper part of the Neapolitan Yellow Tuff deposits, the coastal regressive sands (buried by the post-glacial drape), the submerged or buried parasitic vents or maars (Somma-Vesuvius), the volcanic edifices (tuff cones, domes and their remnants), the remnants of the Late Pleistocene continental shelf have been distinguished and mapped based on the Multibeam and Chirp analysis. During the Cretaceous the outcrops of carbonatic acoustic basement, the landforms with shelf break, the terrace rims, the canyon walls, the slides and scars along canyon walls, the slump scars, the gravity slumps, the drainage patterns, the "Banco di Fuori" slope, the "Ammontatura" channel rim and the debris flow outrunner blocks have been distinguished.

In the Naples Bay the continental shelf has a variable width, ranging from 2.5 km, offshore the western side of the Capri island, to about 10-15 km, offshore the Sorrento Peninsula. On the contrary, offshore the Vesuvius volcano the morphological break, which normally separates the continental shelf from the upper slope, is not clearly identified. Offshore the Posillipo hill a morphological break occurs near the 90 m isobath. Such a complex physiography is accounted for the interference of subaerial and submarine volcanism, coupled to tectonism and linear erosion operated by the Dohrn and Magnaghi canyons and their tributary channels.

The Naples Bay continental shelf is covered by a recent (<10 ky) drape; most of the seabed features have been recognized through the seismic interpretation in Holocene sediments; where the resolution of profiles does not allow the identification of the Holocene drape, the analysis of bottom samples (box-cores) has confirmed the wide occurrence of a thin veener of recent sediments.

The geological interpretation of Multibeam bathymetric data has shown the occurrence of two morpho-bathymetric highs, sedimentary in nature, located next to the heads of the Dohrn canyon at the center of the bay. Seismic stratigraphy has revealed that they consist of deltaic sediments arranged with clinoform patterns and related to the distal part of a fossil prograding wedge fed by the paleo-Sarno river. These remnants have an elevation with respect to the surrounding average depth, that is in the order of 20-30 m. This anomaly can be accounted for the differential erosion, occurred during the shelf subaerial exposure related to eustatic lowstand. Seismic evidence allows to infer a pre-Campanian Ignimbrite age (e.g. 35 ky) of this erosional phase.

In the Torre del Greco offshore area submerged or buried parasitic vents, genetically related to the Somma-Vesuvius volcano, are concentrated along a NE-SW trending. A tectonic lineament, showing evidence of normal faulting, has been previously suggested in the Somma-Vesuvius area (Finetti and Morelli, 1974), based on the interpretation of multichannel seismic profiles. Next papers on the geology of Somma-Vesuvius offshore have inferred different tectonic lineaments, offshore the Somma-Vesuvius volcanic complex (Aiello et al., 2010). A three-dimensional reconstruction of a large volcanic structure located offshore the town of Torre del Greco (Somma-Vesuvius offshore) and representing the seaward prolongation of the Vesuvius volcano has been carried out using the integrated geological interpretation of existing densely spaced high resolution seismic and magnetic profiles. This evidence is confirmed by the interpretation of seismic profiles, showing three main vertical culminations of the volcanic structure, where the overlying sediment drape is significantly reduced. These culminations are linked to magnetic anomaly extremes, having values ranging between 250-350 nT. A striking coincidence between the long-shaped topographic high on the sea floor (towards the east) and the high in the top of the volcanic structure underneath can be explained as a consequence of sea floor deformation due to the underlying volcanic culminations. Smaller volcanic mounds, buried and fossilized by marine sediments, are not related to any magnetic anomaly, probably due to their composition of volcanic tuffs (Aiello et al., 2010).

4.2. Morpho-Bathymetry and Sea Bottom Instability of the Naples Bay and Southern Ischia Canyons

4.2.1. Naples Bay

An extensive, high resolution bathymetric survey of the continental shelf/slope system of Campania (Southern Italy) has been recently carried out. The relative bathymetric data were acquired during the years 1997 and 2002, using Multibeam systems with an average vertical resolution of < 10 m. The survey data were successively merged with a Digital Terrain Model (DTM) created from topographic maps of the Bay of Naples onshore coastal areas and islands, to produce a Digital Elevation Model (DEM) based on homogeneous grid with cell-spacing of 20 m (Figure 4).

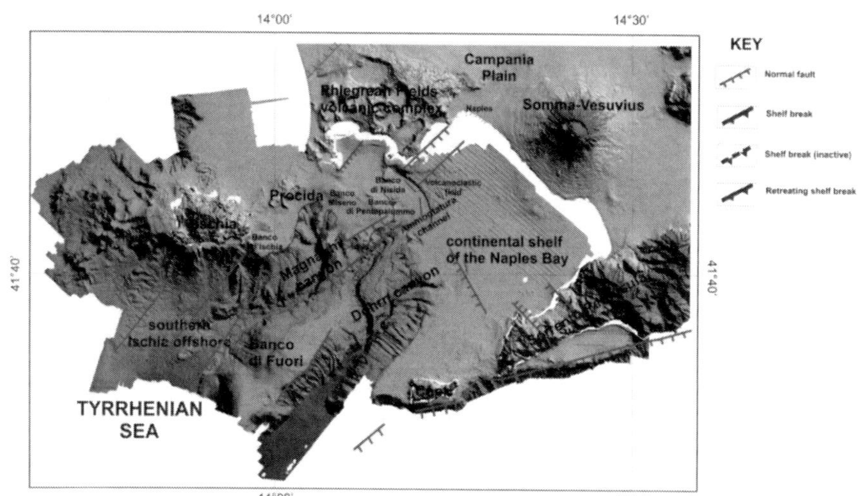

Figure 4. Sketch interpretation of regional morpho-structures on onshore/offshore Digital Elevation Model of the Naples Bay. 1 Normal fault. 2 Regional seismic profile. 3 Shelf break. 4 Shelf break (inactive). 5 Retreating shelf break (modified after Aiello et al., 2011).

The shaded-relief map has make available new and detailed information on the morphology of the Campania coasts. This segment of the Italian continental margin exhibits evidence of the latest Neogene-Quaternary interplays between tectonism and volcanism, as well as of the depositional processes largely developing as a consequence of frequent and large volcano-sedimentary supply. The major morphological features revealed by the 3D-digital maps are: i) the system of the marine canyons (Dohrn and Magnaghi) that cut the continental slope at a depth between 250 m and 1100 m; ii) the continental slope system of the Ischia volcanic structure (Chiocci and de Alteriis, 2006: Aiello et al., 2009a); iii) the onshore and offshore volcanoes of the Phlegrean Fields volcanic complex (Rosi and Sbrana, 1987; D'Argenio et al., 2004; Aiello et al., 2004; 2005; Milia and Torrente, 2012); iv) the rugged seafloor area of the outer shelf of the city of Naples (Banco della Montagna; Sacchi et al., 2000; D'Argenio et al., 2004); v) the debris flow/avalanche deposits on the inner continental shelf off Mt Vesuvius (Milia et al., 1998; 2008), laterally grading into the Torre del Greco volcanic structure (Aiello et al., 2010).

The DEM of the Bay of Naples (D'Argenio et al., 2004) representing the base of the geologic and morpho-bathymetric interpretation, has a 20x20 m cell and has been based on the integration of different grids. This cell has resulted adequate in detecting the most prominent topographic and bathymetric features of the coastal zone, larger than a few tens of meters.

Despite the great number of geological and volcanological studies on the Gulf of Naples, among which Latmiral et al. (1971), Finetti and Morelli (1973), Cinque et al. (1997), Pescatore et al. (1984), Fusi et al. (1991), Milia (1996), Milia et al. (1998; 2008); Aiello et al. (2001; 2004; 2005; 2009a; b; 2010; 2011a; b), Bruno et al. (2003), only recently submarine gravity instabilities in the Dohrn and Magnaghi canyons have been investigated in detail by Di Fiore et al. (2011) and the areas of incipient submarine sliding accurately defined.

The Dohrn canyon formation was triggered off by the tectonic uplift of both the outer shelf and the fluvial valley mouths, during periods of eustatic fall of sea level (Milia, 2000). During the Late Quaternary the Bay of Naples continental slope was characterized by slumping and canyon formation and therefore it may be considered to represent an erosional slope-system (Ross et al., 1994; Galloway, 1998). This implies that megaslides commonly nucleate one or more submarine canyons and that retrogressive slumping is a fundamental process in canyon formation (Coleman et al., 1983; Goodwin and Prior, 1989). This implies that a canyon develops upwards from a base of the slope to the shelf.

The geological conditions for the formation of the Dohrn canyon appear to be incompatible with models based on oversteepening, high sediment supply, sea level rise and retrogressive slumping. The Dohrn canyon developed along a central slope, showing low gradients and the landward rotation of the platform block resulted in a substantial decrease of the slope. Moreover, the canyon wall cuts the main body of the slumps, whereas the most prominent scars show no correlation with the canyon walls. In this way, the Dohrn canyon formation has been controlled by a sea level eustatic fall, inducing a seaward migration of the river systems during the Late Pleistocene and the formation of the valleys, coupled with a fault block rotation, is responsible for the uplift of the outer shelf (Milia, 2000).

According to Canals et al. (2006), the continental slope is a steep, narrow fringe separating the coastal zone from the deep ocean. During sea level lowstands, slides and dense, sediment-laden flows erode the outer continental shelf and the continental slope, leading to the formation of submarine canyons, that funnel large volumes of sediments and organic matter from shallow regions to deep oceans. During sea level highstands, such at present, the canyons have experienced occasional sediment gravity flows, which are usually thought to be triggered by sediment failure or river flooding. The canyon formation has received a considerable debate, but their demise has received less attention. Research on modern canyons with cores and moorings has documented active sediment transport and deposition, but extrapolation of these local observations over larger areas is precluded by complex canyon geomorphology (Walsh et al., 2009). Highstand sedimentation has been recently demonstrated as a possible mechanism for canyon burial and cause of canyon demise, which has important implications for the evolution of canyon globally (Walsh er al., 2009).

On the other side, submarine gullies are small-scale, straight and shallow channels formed in a relatively high seafloor slope settings, playing an important role in the general evolution of the continental margin morphology. The mechanisms associated with the origin and the evolution of submarine gullies are however, still poorly defined. Evidence of topographic signature of gully erosion in the Cook Strait Sector of the Hikurangi subduction margin have been recently detailed by Micallef and Mountjoy (2011). The submarine gully initiation is a threshold process driven by unconfined, directionally stable fluid or sediment gravity flows accelerating downslope. Cascading dense waters, i.e. currents driven by seawater density contrasts, have been proposed as one source of these flows (Micallef and Mountjoy, 2011).

Submarine gullies on Italian upper slopes and their relationships with volcanic activity have been recently revisited (Chiocci and Casalbore, 2011). Multibeam bathymetry and high resolution seismic profiles have been depicted in detail the characteristics of submarine gullies present in the upper continental slope offshore the mouths of Tiber and Volturno rivers in the Tyrrhenian sea and the Simeto river in the Ionian sea. The upper slope gullies have been interpreted as depositional features, growing because of fast aggradation on intergully areas with respect to their axes. The dispersal of river flood sediments through plunging of hyperpicnal flows has been interpreted as the limiting factor of sedimentation in gully areas, also if it requires a sediment concentration of several tens of kilograms in the river flows to overcome the higher density of seawater. This threshold has not been reached by medium-sized rivers, as the Tiber and Volturno rivers. For the Tiber river it has been proposed that, during lowstand phases, a large amount of underconsolidated sediments, deposited during the previous highstand stages, have been eroded from the river valleys, coastal plain valleys and shelf deltas. For the Simeto and Volturno rivers the occurrence of large volcanic edifices (de Alteriis et al., 2006; Aiello et al., 2011a) might be a main control factor, increasing the sediment load in watercourses and favouring the formation of hyperpicnal flows (Chiocci and Casalbore, 2011).

Based on the interpretation of Multibeam bathymetry we have already constructed a sketch geomorphological map of Dohrn and Magnaghi canyons and of the surrounding continental slope, including the southern slope of Ischia volcanic complex, showing, in particular, the slide scars and the canyons'axes have also been drafted. The large slide scar of the southern flank of the Ischia volcanic complex corresponding to the Ischia Debris Avalanche (Chiocci and de Alteriis, 2006; Aiello et al., 2009) has been indicated. The main canyons eroding the southern slope off the Ischia volcanic complex are also shown.

Several bathymetric profiles along the Dohrn-Magnaghi canyon system have also been constructed for a better geological comprehension of the erosional and depositional processes of the Naples continental slope (Figure 5; D'Argenio et al., 2004; Ruggieri et al., 2007). Detailed bathymetric contour maps of Dohrn and Magnaghi canyons reporting the location of bathymetric profiles are also reported (Figure 6; Ruggieri et al., 2007).

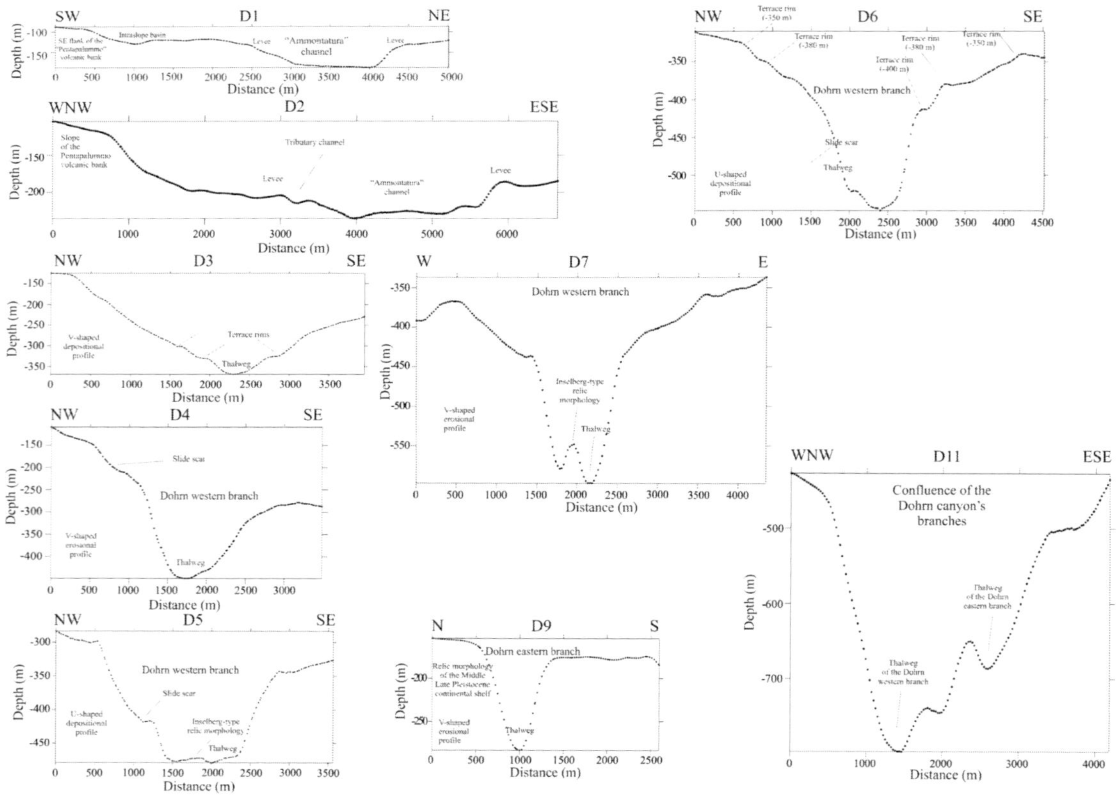

Figure 5. Bathymetric profiles crossing the two branches of the Dohrn canyon. Schematic geologic interpretation is also reported (modified after Ruggieri et al., 2007).

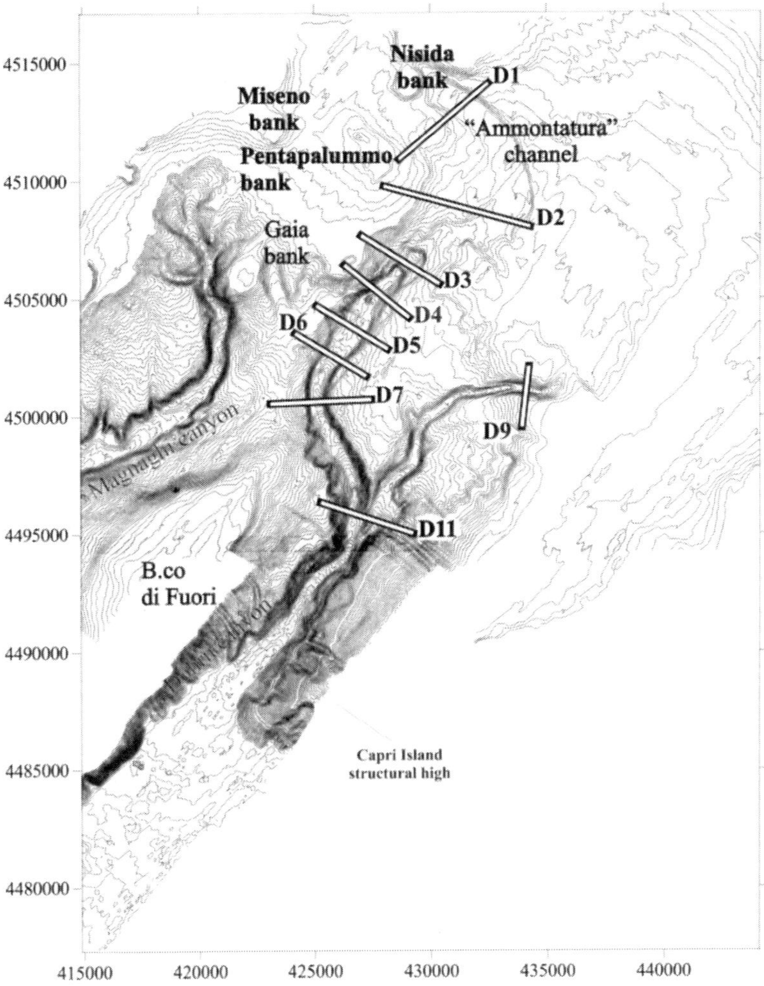

Figure 6. Multibeam bathymetric contour map of the Dohrn canyon. The location of bathymetric profiles shown in Figure 5 is also reported (modified after Ruggieri et al., 2007).

The Dohrn canyon's width ranges from a few hundred meters to more than 1 km, its depth from 250 m at the shelf edge to some 1300 m at the merging with the bathyal plain: the dip of its walls attains some 35° in the steepest sectors. This canyon starts with two major curved branches. The western branch merges into the shelf through a 1.5 km wide and 20-40 m deep channel ("Ammontatura Channel), characterized by a flat bottom and asymmetrical levees, located along the – 200 m isobath and by a sinuous shape in plan view. The Dohrn eastern branch shows a meandering trend and starts from the shelf break of the Sorrento Peninsula, located along the – 120 m isobath. The Dohrn western branch is broader than the eastern one and more deeply incised; the two branches form a typical Y-structure.

The abrupt termination of the Ammontatura shallow channel against the Nisida volcanic bank, whose growth was older or contemporaneous with the eruption of the volcanic deposits of the NYT (Neapolitan Yellow Tuff; Orsi et al., 1992; 1995; Scarpati et al., 1993; Pappalardo et al., 1999; Deino et al., 2004; D'Antonio et al., 2007; Pebst et al., 2008) suggests that the Dohrn canyon system is older than the volcanic deposits of the NYT, which forms the main basis of the city of Naples. This implies that most part of the canyon system activity was probably older than 11-12 ky B.P., age of the NYT deposits in the city of Naples (Di Girolamo et al., 1984; Rosi and Sbrana, 1987).

The Magnaghi canyon shows a triple incised head and is pervasively affected by small-scale slope instabilities. It runs parallel to the southern flanks of the Procida and Ischia islands and then is buried below the Ischia Debris Avalanche (Chiocci and de Alteriis, 2006; Aiello et al., 2009a). Three main tributary channels join basinwards into the main axis. The erosion and transport of volcanoclastics in the western sector of the gulf, near the Ischia and Procida islands, developed along the Magnaghi canyon axis and it appears unrelated to the present fluvial drainage system on land.

As already noted, the Bay of Naples canyons start from the shelf break near the Phlegrean Fields volcanic district, located along the – 140 m isobath. Their trending is controlled by the morpho-structures of the Bay (Figure 4): the Banco di Fuori structural high, bounding southwards the whole canyon system (Aiello et al., 2011b) and the Capri structural high, bounding eastwards the Dohrn canyon, near its confluence with the bathyal plain. Both the Magnaghi canyon and the Dohrn western branch show morphological evidence of retreating of the canyon's head.

The bathymetric profiles located along the Dohrn western branch confirmed the occurrence of several slide scars, already evidenced in plan view by the Multibeam bathymetry. Rounded morphologies, located in the canyon's thalweg are interpreted as relict structures, probably due to a selective erosion acting along the canyon's valley (bathymetric profiles D5-D7). The section D3, carried out in the retreating Dohrn canyon's head, shows terrace rims respectively located at 340 m and 300 m of water depth. This morphological evidence suggests at least two phases in the activity and retreat of the canyon's head. Well-developed terrace rims (at least five rims) have been recognized also on the bathymetric profile D6, respectively located at water depths of 520 m, 400 m, 380 m, 350 m and 320 m. They suggest the occurrence of several phases of erosion and deposition in the canyon evolution.

The geological interpretation of bathymetric profiles allowed us to study the distribution of the V-shaped erosional profiles with respect to the U-shaped depositional profiles in the whole Dohrn canyon system. A relative abundance of V-shaped erosional profiles has been observed in both the branches of the canyon (profiles D3, D4, D7; D9, D11). The U-shaped depositional morphologies, suggesting recent phases of canyon filling occur mainly in the wider, southernmost sector of the Dohrn western branch and at the confluence of the two branches.

A dense network of tributary channels, controlling the overflowing of sediments in the surrounding areas of the continental slope, laterally fed the Dohrn eastern branch. Two main tributary channels start from the shelf break of the Phlegrean submarine volcanic banks (located along the 140 m isobath) and run along the continental slope between the Dohrn and Magnaghi canyons, giving rise to channel-lobe systems. A complex system of tributary channels, located at water depths ranging from 200 m and 500 m, fed the Dohrn eastern branch starting from the shelf break off the Sorrento Peninsula; it seems partly fault-controlled, as suggested by the rectilinear shape in plan view of the southernmost fault-channels in Multibeam bathymetry. The hypothesis that these channels are fault-controlled was already proposed after a deep-towed TOBI Sidescan Sonar survey of the sea bottom in the Bay of Naples (Chiocci et al. 1998a). A system of lobes, genetically linked with some of these channels, has been recognized on a morphological terrace located at water depths of 340 m. Fossil tributary channels suspended over the main branches testify phases of rapid re-incision, switching off the feeding from lateral sources and forming suspended valleys.

Submarine slides are controlled by the loading and the underconsolidation of sediments due to rapid sedimentation, oversteepening of slopes consequent to phases of deep linear erosion, earthquakes and sea level changes (Embley et al., 1982; Saxov and Niewenhuis, 1982; Coleman et al., 1983; Gawthorpe et al., 1985; O'Leary and Dobson, 1992; King et al., 1996; Mulder and Cochonat, 1996; Elverhoi et al., 1997; Holmes et al., 1998; Piper et al., 1999; Watts et al., 2001; Bunz et al., 2003; Posamentier and Kolla, 2003; Sultan et al., 2004; Cartwright et al., 2005, Frey-Martinez et al., 2006; Minisini et al., 2007; Aiello et al., 2009a; Van Der Meerwe et al., 2011).

Several submarine slides and scars are still evident on the canyon's walls and on the continental slope as well. The most part of the slide scars involving the Dohrn canyon appear to be concentrated along the western slope of its western branch; the other branch does not appear involved by mass wasting. Slide scars have been identified by geologic interpretation also on bathymetric profiles (profiles D4 and D5).

Submarine mass failure or underwater landslide are broad terms that encompasses reef failure, submarine rock slides, underwater slides and underwater slumps (Schwab et al., 1993). There are no clear definitions of these terms: rocks may be significantly weathered and sediments may be only partially lithified; sands and silts may possess cohesion and cohesive muds may contain some sand and silt. These materials are often found to coexist in stratified layers. Multiple failure mechanisms often act, such as increased pore water pressure combined with a nearby earthquake (Hampton et al., 1996). Different types of submarine mass failures may occur at the same time and may trigger one another (Tappin et al., 1999; 2001). A strong earthquake can trigger thousands of landslides, both subaerial and submarine (Wilson and Keefer, 1985; Watts, 2001).

Based on the interpretation of Multibeam bathymetry the main features that we interpret as slide scars are located in the following areas (Figure 7):

a) on the head of the Dohrn western branch, showing a double regressive head, controlled by extensive submarine erosion;
b) on the western slope of the Dohrn western branch at its boundary with the eastern flank of the Banco di Fuori morpho-structural high (here a set of coalescent, large slide-scars, not related to the canyon's thalweg may be observed);
c) on the continental slope, north of the Capri structural high, where large scars are suggested by the trending of the isobaths next to the Dohrn canyon thalweg.

Here the canyon system is characterized by several terraces and three rectilinear gullies converge into the canyon from the northern slope of Capri island. An acoustic basement, probably composed of Mesozoic-Cenozoic carbonate rocks almost reach the sea bottom, under a thin drape of Holocene sediments.

The geomorphological interpretation and the analysis of canyon morphometric parameters suggest that the submarine instabilities of the canyon system are located (Figure 7):

a) around the Dohrn western branch, from the canyon's head to the middle of the branch, north of the Banco di Fuori morpho-structural high;
b) on the continental slope, southwards of the Magnaghi canyon, where two large areas of instability, not connected with the canyon's thalweg, may be observed;
c) on the north-western slope of the Banco di Fuori morpho-structural high, where the concave trending of the isobaths suggests the occurrence of an incipient and/or fossil slide scar.

No slumped masses are preserved in the main thalweg, suggesting a probable activity of low-density turbidity currents during the Late Quaternary evolution of the canyon. Large isolated remnants, showing average dimensions of 300x400 m across are present on the canyon floor which, in its upper part, is scoured by a minor meandering channel. These rounded morphologies are interpreted as relict structures, probably due to a selective erosion acting along the canyon's valley. Moreover, terrace rims, located respectively at – 340 m and – 300 m of water depth, suggest at least two phases in the activity and retreat of the canyon head.

A relative abundance of the V-shaped erosional profiles has been observed in both the branches of the Dohrn canyon. The flat-bottomed valley depositional morphologies, suggesting recent phases of canyon filling, occur mainly in the wider, southernmost part of the Dohrn western branch and at the confluence of the Dohrn northern and southern branches.

The drainage system of the canyons is composed of a dense network of tributary channels, controlling the overflow of sediments in the surrounding areas of the continental slope. Two main tributary channels originate from the shelf break off the Phlegrean submarine volcanic banks, located along the 140 m isobath and run along the continental slope between the Dohrn and Magnaghi canyons, giving rise to channel-lobe systems. A complex system of tributary channels, located at water depths ranging from – 200 m and – 500 m, fed the Dohrn southern branch, starting from the shelf break off the Sorrento Peninsula and it appears fault-controlled. This interpretation is suggested by the rectilinear shape in plan view of the southernmost four channels. A system of lobes, genetically linked with some of these channels, has been recognized on a morphological terrace located at water depths of 340 m. Moreover, fossil tributary channels hung over the main branches testify stages of rapid re-incision, switching off the feeding from lateral sources and forming suspended valleys.

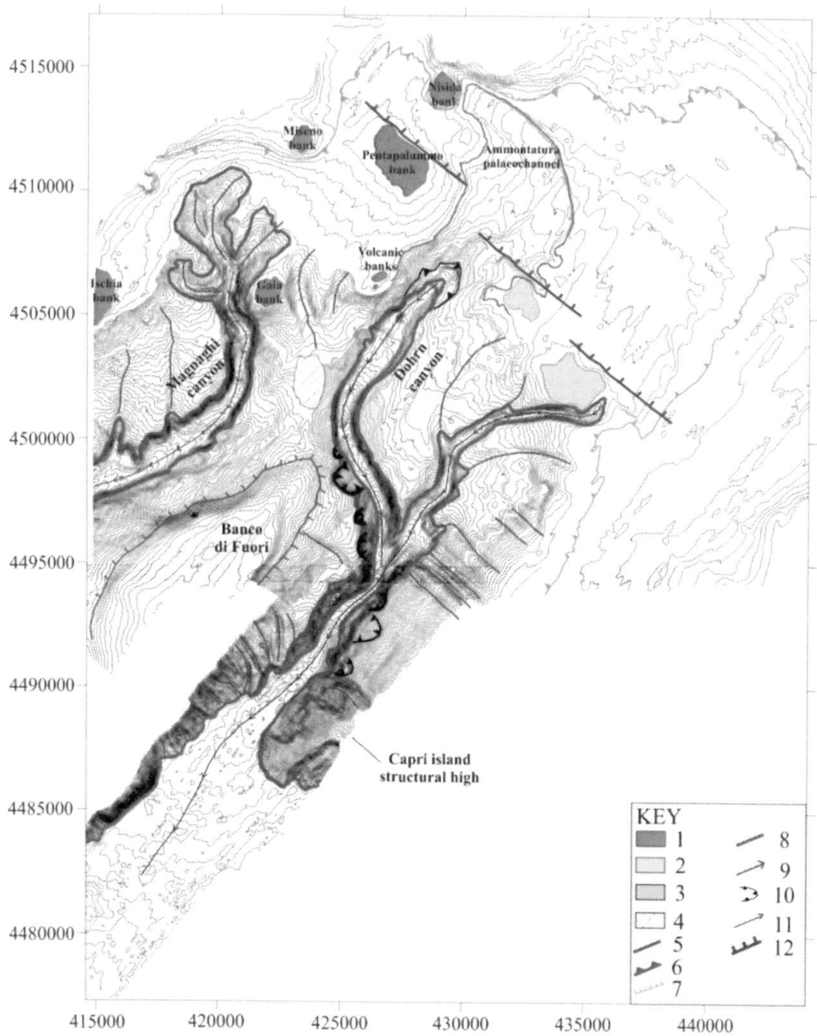

Figure 7. Morpho-bathymetry of the Naples Bay canyons and gravity instability map (modified after Di Fiore et al., 2011). Note that chief submarine instability areas are located at the Dohrn canyon's head, on the slope surrounding the western Dohrn branch, on the slope southwards of the Magnaghi canyon and on the north-eastern slope of the Banco di Fuori structural high. Figures on the side refer to kilometric coordinates.

In conclusion, several submarine slides and scars are evident on the canyon's walls, especially along the western flank of the Dohrn northern branch and on the continental slope as well. The north-western branch of the

Dohrn canyon is affected by instability, with an incipient slump causing a broad depression, 200-300 m across and away from the canyon's edge, and a semi-circular scar on the canyon walls, showing lateral coalescence and defininig a large area of instability, located westward of the north-western slope of the Banco di Fuori morpho-structural high.

4.2.2. Ischia island

The geological interpretation of the marine DEM of Ischia island has allowed for the identification of the main morpho-structural features of the sea bottom. Spectacular features characterize the continental slope off south-western Ischia island, incised by a dense network of canyons and tributary channels, starting from a retreating shelf break, parallel to the coastline and located at varying depths (de Alteriis and Toscano, 2003; de Alteriis et al., 2005; Chiocci and de Alteriis, 2006; Aiello et al., 2009a; b). Large scars characterize the platform margin off south-western Ischia island, in particular the scar of the southern flank of the island, corresponding onshore to the Mt.Epomeo horst and probably at the origin of the Ischia Debris Avalanche (IDA; Chiocci and de Alteriis, 2006). Volcanic banks, having irregular morphologies, have been identified on the south-western flanks of the island, as the "Banco di Capo Grosso", a complex morpho-structural high located on the southern continental slope.

The marine DEM shows the relic morphology of the Banco di Ischia, a wide terraced volcanic bank located on the south-eastern flank of the island. Swath bathymetric surveys of Ischia (de Alteriis et al., 2005; Passaro, 2005) coupled with Sidescan Sonar imagery and seismic profiles put in evidence a large field of hummocky deposits south of Ischia island (*the Ischia Debris Avalanche*). Piston coring coupled to tephrostratigraphy already suggested that the avalanche occurred during prehistorical times (Chiocci and de Alteriis, 2006).

Volcanic structural collapse in the form of avalanches, rock fall and landslides may be almost any size, ranging from a few loose rocks falling from the crater rim of a volcano to large avalanches, such as one at Socompa volcano in Chile, that covers 500 square kilometers (Francis, 1993). Large-scale debris avalanches are related with many volcanic areas, including Augustine Volcano in Alaska, Bandai-san in Japan and Mounts St. Helens in Washington. Submarine landslide deposits have been found off the coast of several Hawaiian islands, including Molakai and Hawaii.

A debris avalanche is formed when an unstable slope collapses and debris are transported away from the slope. Large scale avalanches normally occur on very steep volcanoes. There are two general types of debris avalanches: the cold ones and the hot ones. A cold debris avalanche usually results from a slope becoming unstable, whereas a hot debris avalanche is the result of volcanic activity, such as volcanic earthquakes or the injection of magma triggering slope instabilities. For several months, before Mount St. Helens erupted in 1980, magma was injected into the northern flank of the volcano, creating a bulge that was extremely unstable. An earthquake triggered the movement of land in this area, and the result was a fast moving debris avalanche.

Geological data have shown that a larger a debris avalanche is, the faster it moves. The reason for this is due to energy stored in material within the avalanche. Some large avalanches have been known to carry blocks as large as three kilometers in length, several kilometers from their sources. Such avalanches can travel close to 300 km/hr (Francis, 1993).

In volcanic areas debris avalanche deposits have been defined as the product of a lateral collapse of a volcanic edifice, in conditions of under-saturation of water. The deposit is characterized by two components, respectively called blocks (block facies) and matrix (matrix facies). The block facies represents portions of the collapsed volcanic edifice, in which the textural and structural characteristics of the source area may often be recognized. The blocks are normally involved by a network of fractures (jigsaw cracks). The matrix facies, in which the blocks are enclosed, is constituted of a whole of fine-grained particles, coming from several parts of the collapsed sector. Typical morphologies related to a debris avalanche are an amphitheatre depression, corresponding to a source area (see the lateral collapse calderas) and hummocky topographies in correspondence to the surface of the deposit.

A wide geological literature on debris avalanches and related genetic processes and modeling has been produced (Lipman and Mullineaux, 1981; Ui, 1983; Voight et al., 1983; Fisher et al., 1987; Johnson, 1987; McEwen and Malin, 1989; Sousa and Voight, 1995; Wadge et al., 1995; Day, 1996; Belousov et al., 1999; Elsworth and Day, 1999; Laberg and Vorren, 2000; Erismann and Abele, 2001; Heinrich et al., 2001; Denlinger and Iverson, 2004; Normark et al., 2004; Ward and Day, 2006). A new class of granular landslide models, in which avalanches are simulated with a large number of independent particles moving under the influence of gravitational and centripetal acceleration, has been recently produced (Ward and Day, 2006). The computed method has been applied to the 1980 Mount St. Helen debris avalanche by constraining the initial geometry and structure of the slide mass from geologica data. The landslide simulation has generated a deposit whose extent, thickness, morphological structure and lithological variation closely replicate those observed. The modelled avalanche is kinematically consistent with mapped patterns of bedrock scouring, deposit elevation and seismic records. The success obtained in reproducing the features of the Mount Sant Helens avalanche indicates that the debris-deposit data may be used to determine the kinematic histories of less well-observed landslides.

The geological interpretation of the Ischia Digital Elevation Model reveals several volcanic highs, disposed along two main ridges. The first ridge, NE-SW trending, is about 10 kilometers long and includes several morphological highs located in the south-western Ischia offshore at water depths ranging from 800 m and 100 m (Rittmann Bank; P. Buchner Bank; G. Buchner Bank; Banco di Forio). The top of the Banco di Forio tuff cone occurs at water depths of about 30 m. The second ridge, E-W trending, starts in the north-western offshore Ischia, at water depths of about 600 m (from the Banco Pithecusa and the Banco Mazzella) and continues up to the coastal sectors of the island.

The submerged sectors of Ischia island are the site of submarine gravity instability processes, having catastrophic (instantaneous) character (debris avalanches) or continuous character (i.e. accelerated erosion along submarine canyons or channels, debris fluxes along channels and creeping).

The first category of submarine instabilities includes debris avalanches all originated from the volcano-tectonic uplift of the Mount Epomeo block during the last 30 ky. The most important debris avalanche is the IDA (Ischia Debris Avalanche), having a southern dispersal axis with a transport of blocks up to 40-50 kilometres from the island. The origin of the catastrophic collapse of the IDA has to be attributed to a land-sea collapse involving all the southern sector of the island. This is suggested by the large scar of the southern flank of the island, well evident on Multibeam bathymetry and coincident with the Mount Epomeo block.

A morphological sketch map, based on the geological interpretation of Multibeam bathymetry, was drafted to show the main morpho-structural lineaments occurring around the island (Aiello et al., 2009a; 2012a). Three important areas of accumulation of debris avalanche deposits are located in the north, west and south offshore Ischia (Figure 8; Aiello et al. 2009a; 2012a).

The main heads of the canyon are located in the north-western sector of Ischia (Testata di Punta Cornacchia; Canalone di Forio; Figure 8; Aiello et al. 2009a; 2012a). A strong erosion along canyons and channels occurs in the southern offshore, characterized by abrupt slopes, mainly incised in volcanic deposits. Several channels erode the slope between Punta Imperatore and Punta S. Angelo, starting from the retreating shelf break, located at water depths of about 100 m. The preferential NE-SW trending of the channels indicates a possible structural control over their development.

The canyons off the Maronti Bay exhibit a dominant N-S azimuth and start in correspondence to embayments of the shelf break, more retreated than in the adjacent sector of the southern offshore part. The promontories of the platform margin are located in correspondence to promontories of Punta Imperatore, Punta S. Angelo, Capo Grosso and Punta San Pancrazio, where the continental shelf is very reduced or absent.

In the southern offshore Ischia the geological interpretation of the DEM has revealed the reduced extension of the continental shelf, the lobate heads of the canyons off Maronti, Capo Grosso and Barano and the linear heads of the canyons off Cava Grado and Punta San Pancrazio. The continental slope off the Maronti Bay appears to be further retreated and steeper than in the adjacent sectors; this is probably due to the development, to the south, of the large scar corresponding to the IDA.

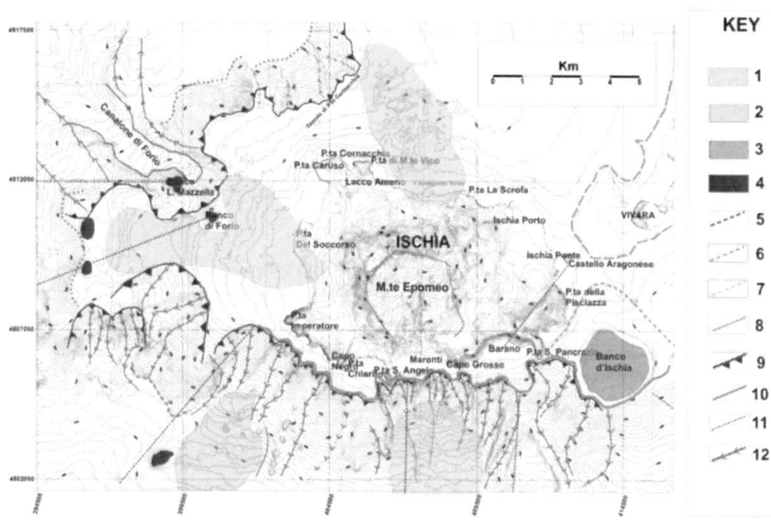

Figure 8. Sketch morphological map of Ischia island, based on the geological interpretation of Multibeam bathymetry. Main morpho-structural features have been represented. 1 slope fan; 2 creeping; 3 hummocky facies corresponding to debris avalanche deposits; 4 volcanic edifice; 5 calderic rim; 6 depositional terrace rim; 7 slope break; 8 channel levee; 9 retreating canyon head; 10 retreating shelf break; 11 buried normal fault; 12 drainage axis; 13 location of seismic profile.

The estimated volume of the IDA, based on Multibeam bathymetry is in the order of 1.5 cubic kilometers. Such a value is coherent with the values estimated on the basis of seismic reflection profiles and piston cores (Chiocci and de Alteriis, 2006; de Alteriis et al., 2010).

The stratigraphic framework for the last 23-ky marine record of the Ischia offshore based on AMC ^{14}C dating and tephrostratigraphic analysis of gravity cores has been recently investigated (de Alteriis et al., 2010). Two collapse events have been recovered in the marine stratigraphic record, i.e. the DA/DF1 deposit, dated between 3 ky B.P. and 2.4 ky B.P. (Holocene deposits) and a former, pre-Holocene DA/DF older than 23 ky B.P. Three ash layers recognized in the post DA/DF1 marine sequence have been correlated to Ischian eruptions occurred between Middle Ages and Roman times; two tephras recovered in the pre-DF1 succession have been correlated with the explosive activity occurred on Ischia and Procida islands from 23 to 17.5 ky B.P. (de Alteriis et al., 2010). This chronostratigraphic reconstruction improved the knowledge of eruptive and collapse events affecting the Ischia offshore during Late Pleistocene-Holocene and their dispersal at sea.

In the northern and western offshore Ischia island hummocky topographies occur, not correlated with evident slide scars. In particular, Sidescan Sonar profiles recorded in western Ischia offshore between Punta del Soccorso and Punta Imperatore show extended hummocky, chaotic facies at the sea bottom, having a kilometric extension at water depths ranging between – 30 m and more than – 100 m. In the hummocky topographies, heterometric blocks are buried and inglobated by pelitic sediments. The deposits reach the tuff cone of the "Banco di Forio" and are interposed between a submerged depositional terrace and a more recent coastal wedge of marine sediments (Budillon et al., 2003). In the northern offshore Ischia hummocky deposits, composed of large blocks, have a fan-shaped topography in plan view and cover a large erosional area offshore Punta della Cornacchia to the west and offshore Casamicciola to the east.

The hummocky area off northern Ischia is evident on shaded relief maps of Multibeam bathymetry (Figure 9). The extension of the deposit is shown in the corresponding figure. The hummocky topographies are located in water depths ranging from – 20 m and – 180 m and show large blocks, having average dimensions in the order of hundred of meters, cropping out at the sea bottom or partly covered by Holocene sediments. The deposits form a large field having sub-circular shape in plan view, which is bounded by depositional levees; a network of erosional gullies develops in the portion of the sea bottom occupied by the hummocks. As suggested by the interpretation of seismic profiles (Aiello et al., 2009a; b) are interpreted as the result of two events of debris avalanches, partly superimposed.

By concluding, the debris avalanches, occurring on the northern, western and southern submerged flanks of the Ischia island are mainly controlled by volcano-tectonic uplift of the Mount Epomeo block, involved in calderic resurgence during the last 30 ky (Orsi et al., 1991; Acocella and Funiciello, 1999; De Vita et al., 2006). The most important debris avalanche of Ischia island is the IDA, having a southern dispersal axis, transporting blocks up to 40-50 km away from the island. The origin of this event has to be attributed to a land-sea catastrophic collapse, involving the southern flank of the island. This evidence is confirmed by the large scar of southern Ischia, well evident on the DEM. The volumetric evaluation of the IDA, based on the DEM analysis furnished values in the order of 1.5 km^3, coherent with the values estimated on the basis of seismic profiles and piston cores (Chiocci and de Alteriis, 2006; de Alteriis et al., 2010).

Differently from the IDA, the hummocky facies, occurring on the western flank of the island (Budillon et al., 2003) and on the northern flank off the Casamicciola harbour, do not appear related to evident slide scars. The complex topography of the sea bottom on Multibeam bathymetry and Sidescan Sonar imagery highlights the occurrence of heterometric blocks, that reach extensions of several hundred of meters, having fallen along the slopes and been inglobated in a sandy-silty matrix. The matrix probably derives both from the volcaniclastic deposits, originally deposited on the slopes and from the failure of sediments, previously accumulated in the inner shelf. The debris avalanche deposits are only partly covered by recent marine sediments, put in evidence by the interpretation of Sidescan Sonar photomosaics. Seismic reflection profiles suggest that the deposit occurring off Casamicciola (northern Ischia) may be the result of two distinct superimposed catastrophic events.

Numerical simulation of the tsunami generated by a past catastrophic landslide in the Ischia island has been recently carried out by Tinti et al. (2011). The consequences of the Ischia Debris Avalanche (IDA) may have caused gigantic sea waves affecting all the coasts of Ischia and of the Gulf of Naples. The IDA and the generated tsunami have been taken as the worst-case scenario for the occurrence of a new tsunami in the area. The simulation of Tinti et al. (2011) has shown that the IDA-induced tsunami attacks severely all the coasts of the Gulf of Naples with the highest waves obtained for the islands of Ischia and Capri and the Sorrento Peninsula. The propagation pattern of the IDA tsunami can be used to get hints on the impact that such an event may have had on early populations habiting Gulf of Naples, but also to get clues that could be most severely hit by a tsunami generated by a smaller-scale landslide that may occur in the same source zone.

In order to elaborate the scenario of the IDA tsunami, the shape of the initial volcanic dome has been reconstructed, based on the hypothesis of a single catastrophic collapse. The landslide motion and the consequent tsunami have been calculated by using the numerical codes UBO-BLOCK1 and UBO-TSUFE, specifically devised for landslide evolution and tsunami propagation (Tinti et al., 2011).

The input data of the geological model include the geometry of the landslide body. This implies the reconstruction of the geometry of the Mount Epomeo in the state prior to the collapse. It has been obtained starting from the present morphology of the volcano, that, being horse-shoe shaped, seems to suggest the failure of a huge mass sector. A full definition of the landslide model geometry has required not only the pre-slide relief of the mountain, but also the reconstruction of the detachment surface of the landslide. Other data required from numerical modelling have included the initial sliding body, the CoM path and the lateral slide boundaries (Tinti et al., 2011).

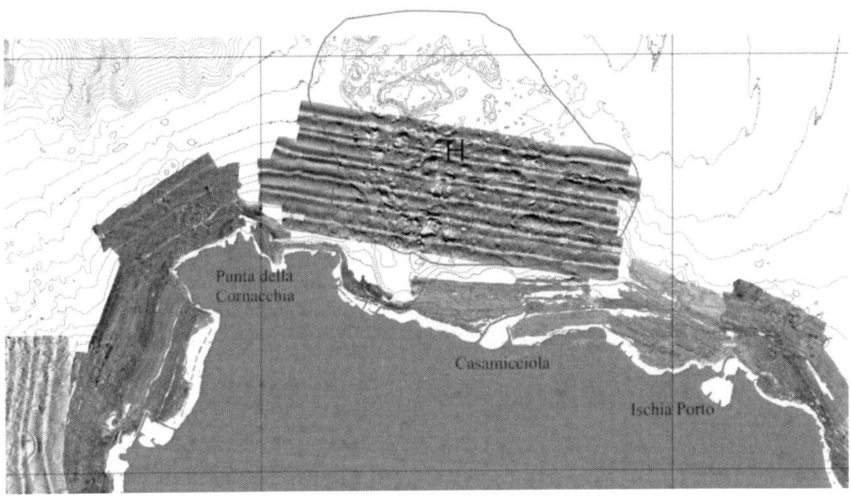

Figure 9. Photomosaic showing the acoustic images around the Ischia island. Note the occurrence of hummocky facies corresponding to debris avalanche deposits in the northern and western offshore of the island.

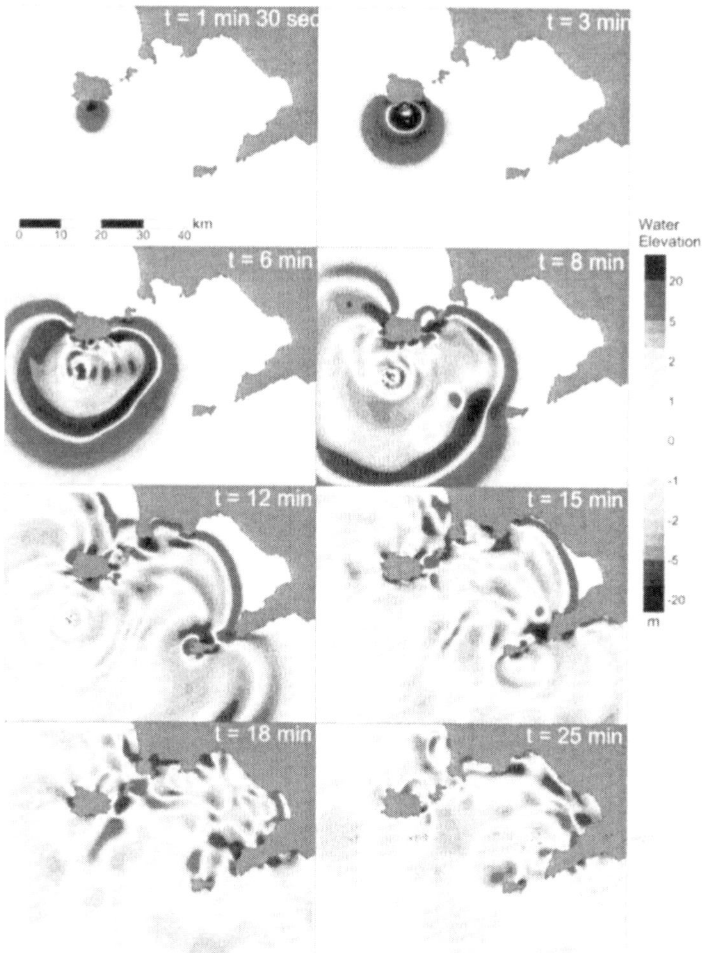

Figure 10. Simulated tsunami propagation in the near field involving Ischia and the Gulf of Naples. Positive sea water elevations are in yellow-red scale, negative elevations in cyan-blue scale (reported after Tinti et al., 2011).

The simulation of the landslide has allowed to deduce that the mean final thickness of the computed deposits ranges between 10 and 50 metres, in agreement with the isopach field deduced from reflection seismic data. The Froude number has also been computed, obtained as the ratio between the horizontal component of the slide speed and the phase velocity of the tsunami wave. It may be taken as the measure of the efficiency of the interaction between the moving mass and the generated wave (Tinti et al., 2011).

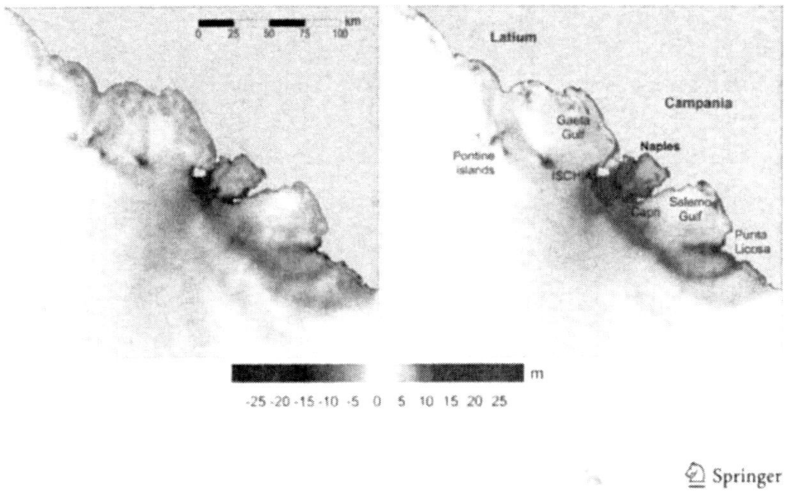

Figure 11. Field of maximum (right) and minimum (left) sea surface elevations of the tsunami induced by the Ischia Debris Avalanche. The computational domain is made of 48.361 nodes and 94.475 triangular elements, embracing an area much broader than the Gulf of Naples that extends up to cover the coasts of Latium to the north, of the entire Campania and northern Calabria (reported after Tinti et al., 2011).

The propagation of the IDA-induced tsunami has been computed over a large grid embracing the coasts of Latium and Campania and is shown through various snapshots taken as different time steps. As shown in Figure 10 (Tinti et al., 2011), after the initial phase, when the wave propagates radially from the source with a positive powerful leading front, the tsunami begins interacting with the nearby islands, namely Procida after 6 minutes and Capri after 8 minutes, with wave amplitudes reaching at least 20 m. Between 8 and 12 minutes the tsunami enters the Gulf of Naples with a strong positive wave, that reaches the town of Naples after about 15 min and the whole coast of the basin in around 18-20 minutes.

The computational domain covers, in the north, the coasts of Latium and of the Gulf of Gaeta and in the south, the Gulf of Salerno, down to Punta Licosa and the Northern Calabrian coast. The maximum and minimum elevation are portrayed in Figure 11 (Tinti et al., 2011) and provide further interesting indications on the tsunami propagation features. A lot of energy is directed towards the island of Capri and the Sorrento Peninsula and more southwards, pointing markedly towards Punta Licosa. The Salerno Gulf is strongly protected by the Sorrento Peninsula. Within the Gulf of Naples, the tsunami heads mainly towards Naples, partially protected by the peninsula where Capo Posillipo is located. With regard to the northward propagation of the tsunami, the Pontine islands are strongly affected by the waves, as a result of the refraction governed by the bathymetry, while the Gaeta Gulf turns out to be partially sheltered.

On the Ischia island the tsunami travel time is quite negligible (Tinti et al., 2011). The wave hits Capri after about 8 minutes, with a positive crest. The wave amplitudes are really impressing in the island of Ischia, exceeding 40 meters, with two or three 20 m peaks affecting the coast in the first 20 minutes. Capri is also reached by relevant waves due to the strong tsunami energy beam determined by the bathymetry. This leads to 40 m maximum waves and a series of 20 m waves, a typical example of tsunami energy traping along the coast.

The evaluation of the tsunami geohazard at a local scale in the case history of the Naples town has been recently evaluated (Alberico et al., 2013). For many years the tsunami have been considered as secondary hazardous events included in the catalogues of the earthquakes (Mallet, 1852-1854; Baratta, 1897) and of the volcanic eruptions (Mercalli, 1883; Baratta, 1897). Only in the next years, with the European projects GITEC (1992-1995) and GITEC TWO (1998-2001) an European catalogue of the tsunami has been published, concealing the Storegga slide, dated back at about 6 ky B.P., up to the 1998 (Tinti et al., 2004). The comparison of the data reported in the catalogue of Tinti et al. (2004) with those recently published by Pasaric et al. (2012) and available on the NOOA website has allowed to verify that in the Mediterranean starting from the 2000 A.C. since to today 467 tsunamis occurred, 157 of which (33%) have involved the Italy.

The earthquakes and the volcanic eruptions are the predominant causes of the tsunami events verified in the Mediterranean. In fact, 376 events have a seismic origin and 18 events a volcanic origin. The same statistic consideration is valid for the Italy with 118 events triggered by seismic events and 12 by volcanic eruptions (Alberico et al., 2013).

These natural events represent a source of risk for the coastal areas of the Mediterranean, hosting about 160 millions of persons and important productive realities. To know the areas which may be hosted in the next period by the tsunamis is of primary importance, both for the hazard prevision and mitigation. To this aim, it has been elaborated a method of a deterministic type for detailed studies feeding the potentiality of the matrix analysis, available in a GIS environment (Geographic Information System) to define the extension of the flooded area and the high of the wave in each point of the land involved by the event.

The input data necessary in such a method (Alberico et al., 2013) are the Digital Terrain Model (DEM) valid for environmental studies and the run-up values of the wave along the coast. The slopes desumed from the DEM are joined in classes and used to calculate the extension of the flooded areas and the new value of the wave height to any slope variation shown by the territory. This method has been applied to the test area of the city of Naples (Gulf of Naples; Alberico et al., 2013), since it is characterized by an articulated morphology allowing to comprehend the role of the topographic slope on the delimitation of the potentially flooded areas.

5. Seismo-Stratigraphic Results

5.1. General Setting

In the Naples Bay the continental shelf has a variable width ranging from 2.5 km (off the western side of Capri island) to about 10-15 km (off the Sorrento Peninsula coast); offshore the Vesuvian area the morphological break, which normally separates the continental shelf from the upper slope, is not clearly identified, while offshore the Posillipo hill a morphological break occurs near the 90 m isobath. Such a complex physiography is accounted for the interference of subaerial and submarine volcanism coupled to tectonism and linear erosion operated by the Dohrn and Magnaghi canyons and their tributary channels.

The whole continental shelf is covered by recent (<10 ky) drape and most of the seabed features have been recognized through seismic interpretation in Holocene sediments; where the resolution of profiles does not allow the identification of the Holocene drape, the analysis of bottom samples (box-cores) has confirmed the wide occurrence of a thin veener of recent sediments.

Two sedimentary bathymetric highs located next to the heads of the Dohrn canyon occur at the centre of the Naples Bay. Seismic stratigraphy has revealed that they consist of deltaic sediments arranged with clinoform patterns and related to the distal part of a fossil prograding wedge fed by the palaeo-Sarno river. These remnants have an elevation with respect to the surrounding average depth that is in the order of 20-30 m. This anomaly can be accounted for the differential erosion that has occurred during the shelf subaerial exposure related to eustatic lowstand. Seismic evidence allows to infer a pre-Campanian Ignimbrite age (35 ky B.P.) of this erosional phase.

In the Torre del Greco offshore area submerged or buried parasitic vents, genetically related to the Somma-Vesuvius volcano, appear concentrated along a NE-SW trend. A tectonic lineament, showing evidence of normal faulting, has been previously suggested (Finetti and Morelli, 1973) in the Somma-Vesuvius area, based on the interpretation of multichannel seismic profiles. In the offshore region between Torre del Greco and Napoli the occurrence of a Holocene unit, rapidly thinning seawards and characterized by chaotic reworked sediments, suggest the emplacement of either lahar-type deposits due to eruptive episodes or mud flows (Figure 12; Aiello et al., 2001). This unit is bordered to the west by mound-shaped highs with an acoustically-transparent seismic facies (Figure 13; Aiello et al., 2001), cropping out at the sea bottom or partially buried by sediments. These morphological highs have been sampled by coring and consist of shaley mounds, with poorly sorted mid to coarse volcanogenic sands and thick pumice levels.

Offshore the Sarno river mouth, at the south-eastern corner of the bay, a typical estuarine sedimentation is present, producing a thick Holocene wedge; creep movements at the sea bottom, probably controlled by horizons with high contents of organic matter, involve Holocene sediments above a sharp surface of separation, which may correspond to the last maximum flooding surface (Capotondi et al., 1994), formed 5000-6000 years ago. In the area offshore of Sorrento-Punta Campanella, the Holocene wedge tapers towards the SW and the Subbottom Chirp profiles show the occurrence of coastal regressive sands covered by a thin Holocene drape, sometimes cropping out at the sea bottom as relic sandy ridges.

The continental shelf south of the Punta Campanella-Capri alignment (in the Salerno Gulf) has not been affected by significant vertical movements from the 5e isotopic substage (Cinque and Romano, 1990) and glacio-eustatic sea level fluctuations represent the main factors controlling the sea bottom morpho-evolution. On the contrary, subsidence occurred north of this alignment, at the shelf margin, since evidence of subaerial erosion is found at depths of 140-150 m. Algal patch reefs, coastal dunes, as well as submerged depositional terraces, often characterize this area and seem to be related to the last glacio-eustatic sea level rise.

A mathematical model for wave and run-up generated submarine landslides in the canyons of the Bay of Naples has recently been purposed (Di Fiore et al., 2011). The results of simulation of failures in the above source areas show that the amplitude of wave run-up, expressed in terms of the sea floor depth percentage, may range up to 2.5% of the water depth at the sea bottom. The morphometric analysis of the Dohrn canyon slope provides insights into tsunami hazard, including the location of mass movements, the size of mass failures, their relative importance for the structure of a continental margin and the potential for landslide-generated tsunami hazard. We do not have enough evidence to consider all these detachment areas located on the Dohrn canyon edges as generated by landslides with impulsive character, or if there has been an evolution over time controlled by small detachments, not significant if compared to the potential needed to generate anomalous waves. The nature of the materials that characterize the area is both volcanic (Campanian Ignimbrite) and sedimentary. Therefore, we deal with a non-coherent material, that, together with the presence of the active volcanoes (Somma-Vesuvius and Phlegrean Fields) needs to be analyzed with regard to its ability to trigger landslides. Moreover, our study suggests that the tsunami hazard is higher on the canyon margin, such several slide scars and large submarine instability areas occur near to the canyon borders. The morphological evolution of the study area depends on complex systems where volcanism, sedimentation and tectonics affect the coast and the marine bottoms, that show the potential to develop slope instability, even though high urbanization has impacted significantly on the sedimentation, reducing the accumulation rate of materials potentially prone to slide (Di Fiore et al., 2011).

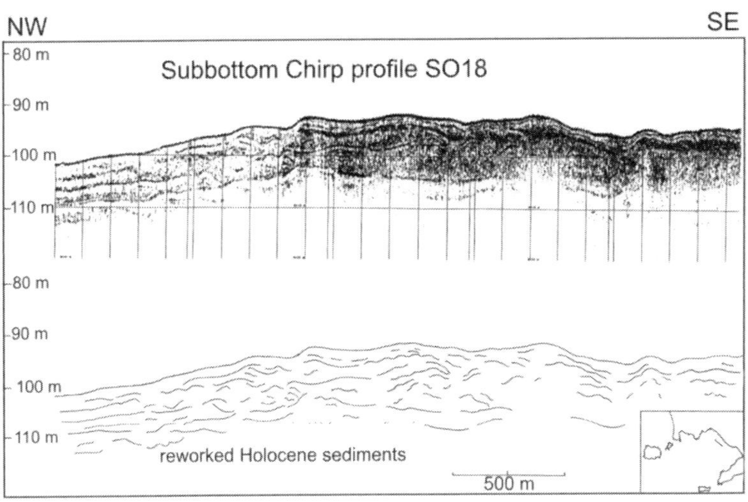

Figure 12. Subbottom Chirp profile SO18, showing the chaotic reflections interpreted as Holocene reworked sediments, genetically linked to the activity of the Somma-Vesuvius stratovolcano and consisting of mud flows and lahars (modified after Aiello et al., 2001).

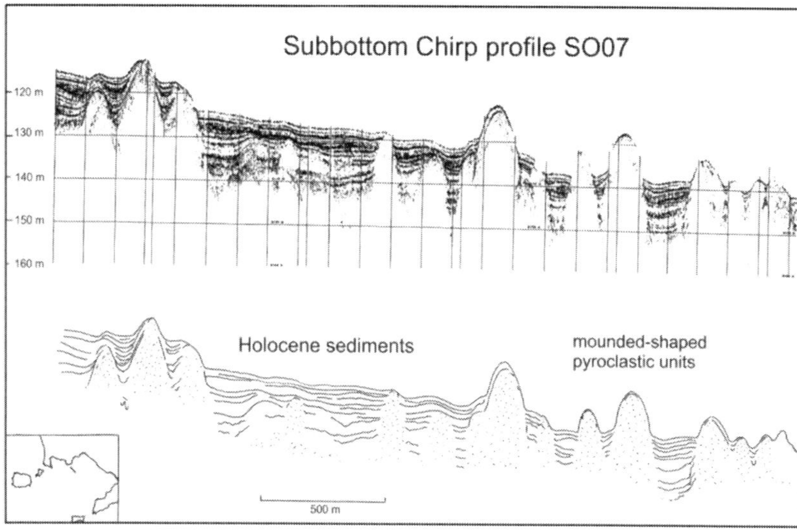

Figure 13. Subbottom Chirp profile SO07 showing mounded shaped morphological highs genetically related to the upper part of the seismic unit interpreted as the Neapolitan Yellow Tuff (modified after Aiello et al., 2001).

5.2. Eastern Gulf of Naples

Selected examples of interpreted seismic and magnetic profiles in the eastern sector of the Gulf of Naples are here discussed to improve the correlation of some volcanic features with corresponding magnetic anomalies (Aiello et al., 2005).

The seismic profile GRNA09 (Figure 14; Aiello et al., 2005) runs from the offshore of Somma-Vesuvius volcanic complex (starting from Torre del Greco town) towards the Sarno river plain with a NW-SE trending. It is located in the inner continental shelf, at water depths less than – 100 m; consequently, the penetration of the single-channel profile is not excellent because of the occurrence at shallow depths of the sea bottom multiple reflection.

The first seismic unit runs from the offshore of Somma-Vesuvius volcanic complex (starting from the Torre del Greco town) towards the Sarno river plain with a NW-SE trending (Figure 14). It is located in the inner continental shelf, at water depths less than 100 m; consequently, the penetration of the single-channel profile is not excellent, due to the occurrence of the sea bottom multiple reflection at shallow depths.

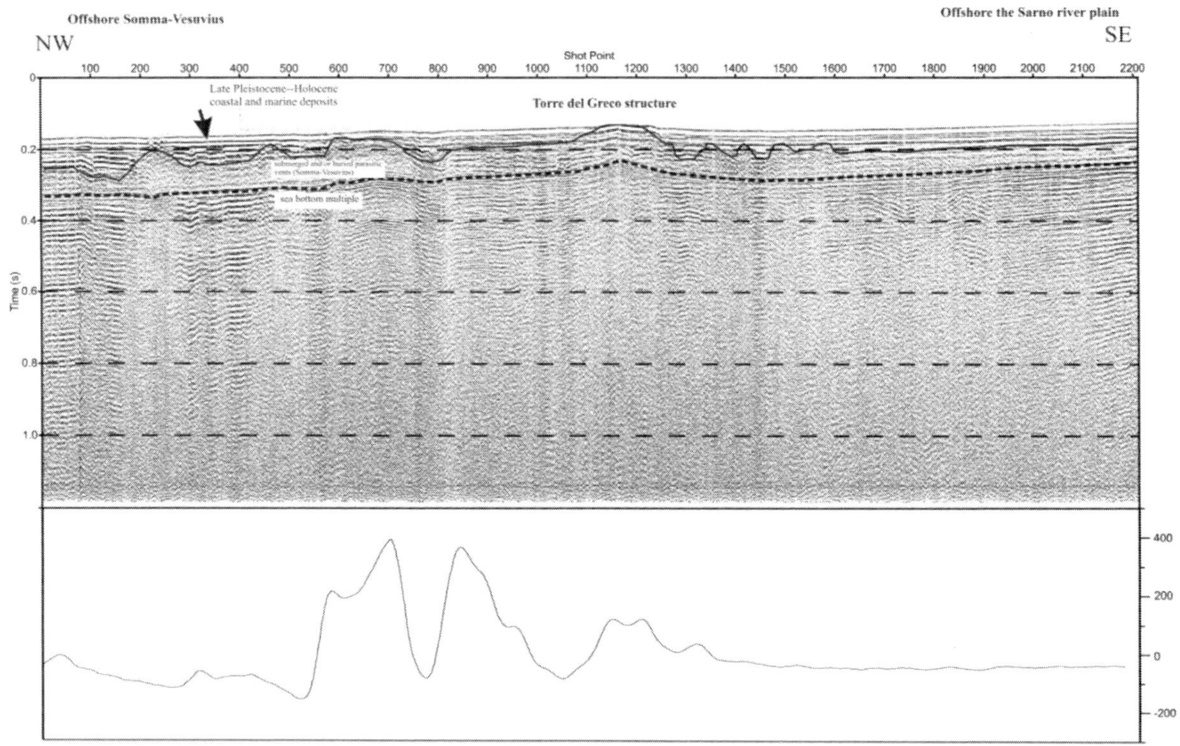

Figure 14. Seismic profile GRNA09 and corresponding interpretation (modified after Aiello et al., 2005). The relative magnetic profile is reported in the inset. Note the occurrence of an important magnetic anomaly (in the order of 500 nT) in a wide volcanic structure located offshore of the Torre del Greco town and related to the Somma-Vesuvius volcanic complex.

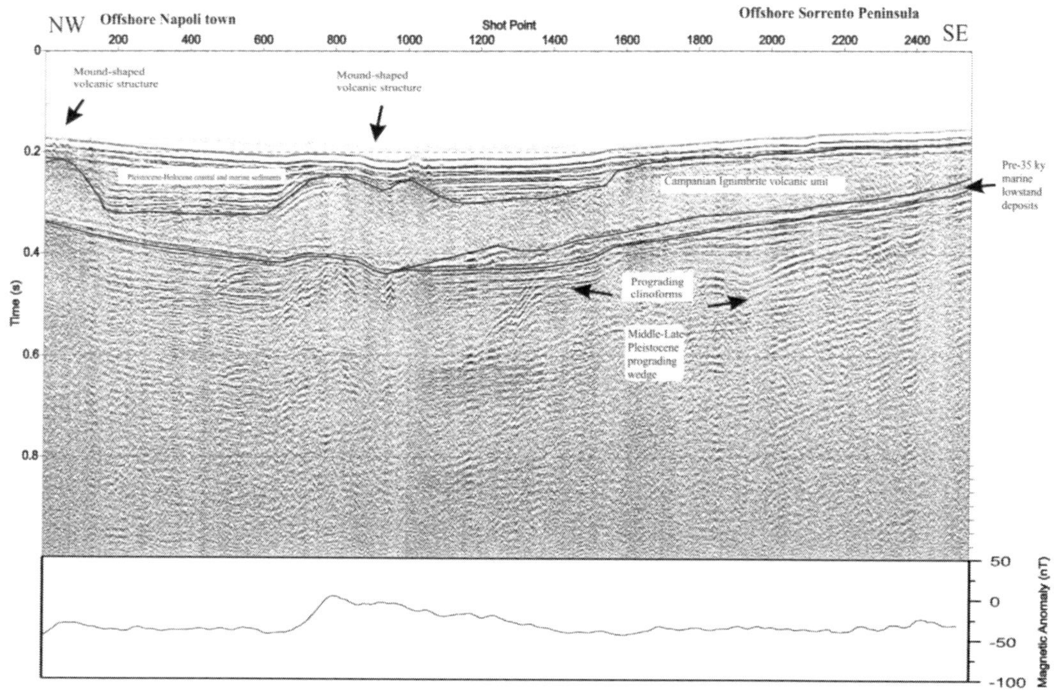

Figure 15. Seismic profile GRNA15 and corresponding interpretation (modified after Aiello et al., 2005). The relative magnetic profile is reported in the inset. Two large, mound-shaped, acoustically-transparent volcanic structures (see the left on the profile) warp the top of a seismic unit correlated to the Campanian Ignimbrite pyroclastic flow deposits (35 ky B.P.). The age of the domes could be very recent, nevertheless they are fossilized by about 300 m of Holocene sediments. A magnetic anomaly with a relative intensity of 40 nT confirms the volcanic nature of these structures.

The first seismic unit is characterized by highly continuous and parallel seismic reflectors, interpreted as Late Pleistocene and Holocene coastal and marine sediments supplied by the Sarno river mouth. The sediments overlie an acoustically transparent volcanic unit, characterized by a tabular external shape; the upper boundary of the seismic unit is highly irregular and runs along the top of volcanic mound-shaped structures. The latter seismic unit is interpreted as the Campanian Ignimbrite pyroclastic flow deposits (Rosi and Sbrana, 1987); the dome-shaped structures warping the Campanian Ignimbrite are interpreted as submerged and/or buried parasitic vents, genetically related to the Somma-Vesuvius volcanic complex (Vezzoli, 1988). These volcanic features were already singled out by Milia et al. (1998) in their paper on Somma-Vesuvius offshore volcanology and by Aiello et al. (2001) in their geomorphological map based on the interpretation of high resolution Subbottom Chirp profiles. The widest of these vents extends for about 7.5 km offshore of the Torre del Greco town; its top shows three main peaks, correlating to the maximum values of the recognized anomalies. The above mentioned structures correspond to a magnetic anomaly field composed of several anomalies, reaching maximum values of 500 nT. This represents one of the maximum values of the observed anomalies in the Gulf of Naples.

The seismic profile GR15 runs in a NW-SE direction offshore of Napoli town up to the Sorrento Peninsula offshore (Figure 15; Aiello et al., 2005). Seismic stratigraphic interpretation shows three units (Pleistocene-Holocene coastal and marine sediments, Campanian Ignimbrite volcanic unit and Middle-Late Pleistocene prograding wedge). A package of parallel reflectors with high continuity and amplitude, interposed between the prograding wedge and the Campanian Ignimbrite is interpreted as pre-35 kyr lowstand marine deposits (Milia, 1996). This unit is not seen at water depth less than 140 m: thus, it was beneath the sea throughout the last major glacial maximum and lowering of sea level.

Large volcanic, acoustically-transparent, dome-shaped structures, were recognized near the top of the Campanian Ignimbrite unit, fossilized by Pleistocene and Holocene coastal and marine sediments and warping the Campanian Ignimbrite unit. The structures are similar to those previously described on the seismic profile GRNA09. The occurrence of a magnetic anomaly, in the order of 40 nT, in one of the dome-shaped structures, confirms its volcanic nature. The age of the domes could be very recent, nevertheless they are fossilized by about 300 m of Holocene sediments. Normal faults bound a basin between the two domes, where sediments significantly thicken.

The same seismic units were recognized on the line GP31 (Figure 16; Aiello et al., 2005), which runs in a NE-SW direction offshore of the Vesuvius area (Torre del Greco town) up to the Gulf of Naples continental slope. The thickness of the Campanian Ignimbrite seismic unit is of about 100 m, comparable with that one of the Campanian Ignimbrite in the "Trecase 1" well, drilled on land in the Somma-Vesuvius area (Balducci et al., 1985). During its deposition the unit probably overlies an inclined palaeomorphology, as evidenced by the landward dipping of the erosional unconformity located at its base, or alternatively, the unconformity was involved by a tectonic tilting of the shelf. Relict morphological highs located in the central part of the gulf, occur near to the present-day shelf break.

The qualitative correlation between the seismic profile and the corresponding magnetic line confirms the volcanic nature of the dome-shaped structure warping the Campanian Ignimbrite unit. The maximum variation of the observed magnetic anomaly is in the order of 120 nT. The intersection of the GP31 and GR15 lines is localized just near to this magnetic anomaly, which is related to the dome-shaped structure identified in both profiles.

The preferential alignment of magnetic anomalies and related seismic structures along a NNW-SSE trending suggest such a structural elongment offshore the volcano (Aiello et al., 2005), never singled out by previous papers which evidenced systems of NE-SW normal faults (Bernabini et al., 1973; Finetti and Morelli, 1974; Cassano and La Torre, 1987; Bruno et al., 1998; Simini et al., 2003; Judenherc and Zollo, 2004). Such a structural elongment is evidenced by the map of the total magnetic field offshore the Somma-Vesuvius volcanic complex. Three intense, dipolar magnetic anomalies are aligned along a direction parallel to the Gulf of Naples shoreline. These anomalies correlate to dome-shaped volcanic structures located along a NNW-SSE elongment in water depths ranging from – 80 m and – 110 m. These structures, located in the outer shelf of the Somma-Vesuvius offshore, are located in some convexities of the isobaths at water depths ranging from – 70 m up to – 110 m.

5.3. Western Gulf of Naples

The seismic profile GRNA51 runs from the offshore of the Phlegrean Fields volcanic complex to the Gulf of Naples continental slope with a NW-SE trending.

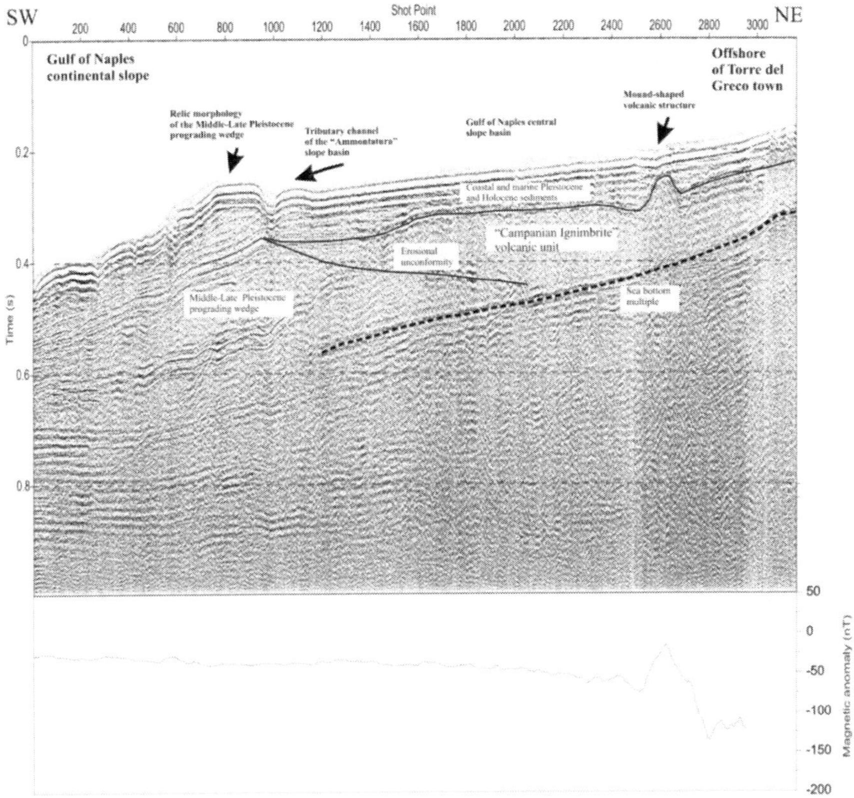

Figure 16. Seismic profile GP31 and corresponding interpretation. The relative magnetic profile is reported in the inset. No magnetic anomaly occurs in the seismic units interpreted as the Middle-Late Pleistocene sedimentary basin filling of the Gulf of Naples continental slope (see the left part of the profile). Offshore the Somma-Vesuvius volcanic complex a magnetic anomaly whose value is in the order of 120 nT corresponds to a dome-shaped volcanic structure warping the Campanian Ignimbrite pyroclastic flow deposits.

In the western part of the gulf (offshore the Phlegrean Fields volcanic complex) the interpretation of the seismic profile GRNA51 has shown the stratigraphic architecture of the outer continental shelf of the Gulf of Pozzuoli area. The Miseno volcanic bank pertains to a belt of volcanic mounds located in the westernmost part of the Gulf of Naples, to the south of the Gulf of Pozzuoli. This submarine belt presents several oval and round-shaped mounds, well evident in the Multibeam bathymetry of the gulf, some of which are characterized by positive magnetic anomalies (Latmiral et al., 1971; Finetti and Morelli, 1974; Pescatore et al., 1984; Fusi et al., 1991). These volcanic submarine features have not been studied in detail, mainly concealing their age and their relationships with the volcanism in the Campania Region.

An acoustically transparent, wedge-shaped seismic unit, interpreted as a volcanic acoustic basement, occurs under the Miseno submarine bank. The excursion of the magnetic anomalies near to the bank is up to 200 nT. This unit grades laterally into a seismic unit characterized by alternating parallel and continuous reflectors with chaotic intervals, showing an erosional truncation near the sea bottom. The last unit could be interpreted as the external part of the Pentapalummo bank volcanic deposits. The shape of the magnetic anomaly in the lateral part of the Pentapalummo bank volcanic deposits reflects the terraced shape of the top of the bank. The excursion of the magnetic anomaly is in the order of about 100 nT. Late Pleistocene marine and coastal progradations, pertaining to the last glacio-eustatic cycle, well developed under the present-day shelf break are located at water depths of about 150 m.

The Dohrn canyon western branch, whose bottom reaches a water depth of about 650 m in this area, erodes volcanic units of its western flank and sedimentary units on its eastern flank (Figure 17; Aiello et al., 2005). A local maximum of the magnetic profile (about 50 nT) is associated with a volcanic intrusion identified along the western slope of the Dohrn western branch. Sedimentary units pertaining to Middle-Late Pleistocene slope sequences of the prograding wedge supplied by the paleo-Sarno river mouth are truncated by both flanks of the canyon. On the bottom of the profile there is an erosional unconformity, representing also the basal unconformity of the Middle Pleistocene prograding wedge, evidenced by the onlap of seismic reflectors.

The seismic profile GRNA85, NW-SE trending, runs from the bathyal plain offshore of the Capri island to the continental slope of the Ischia island. A major morpho-structural high (Banco di Fuori and/or Banco di Bocca Grande) is crossed by the seismic line GRNA85 (Figure 18; Aiello et al., 2005). The Banco di Fuori high separates the Dohrn and Magnaghi canyons and is presumably formed by a Mesozoic carbonate block, resulting from regional uplift and tilting of the carbonatic acoustic basement. The high is covered by Pleistocene-Holocene sediments; its southern flank is steeper than the northern one, thus probably suggesting the occurrence of a normal fault.

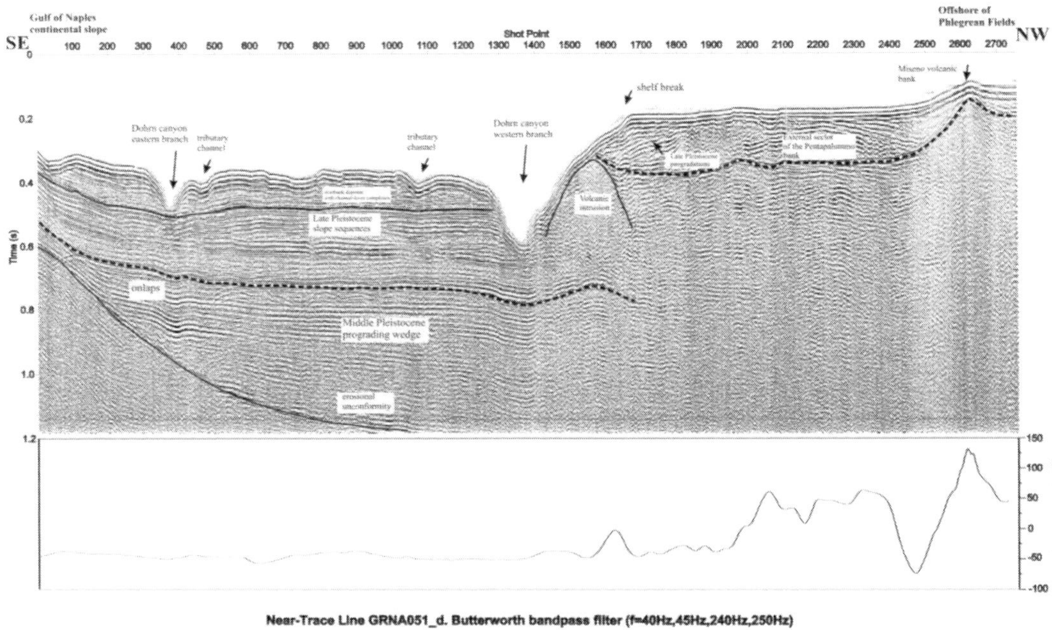

Figure 17. Seismic profile GRNA51 and corresponding interpretation. The relative magnetic profile is reported in the inset. Several main magnetic anomalies have been identified on this profile corresponding to main volcanic bodies offshore the Phlegrean Fields volcanic complex (see on the right in seismic and magnetic profile). A first pronounced anomaly, in the order of 130 nT correlates to the Miseno volcanic bank. A large, terraced anomaly with a relative intensity of 100 nT is related to the Pentapalummo volcanic bank, whose external sector is crossed by the GR51 line. The terraced shape of the anomaly reflects in some way the bathymetric setting of the bank, characterized by a terraced top reaching water depths of 50 m. The flat trending of the magnetic profile confirms the occurrence of sedimentary packages on the Gulf of Naples continental slope, deeply eroded by the two branches of the Dohrn canyon.

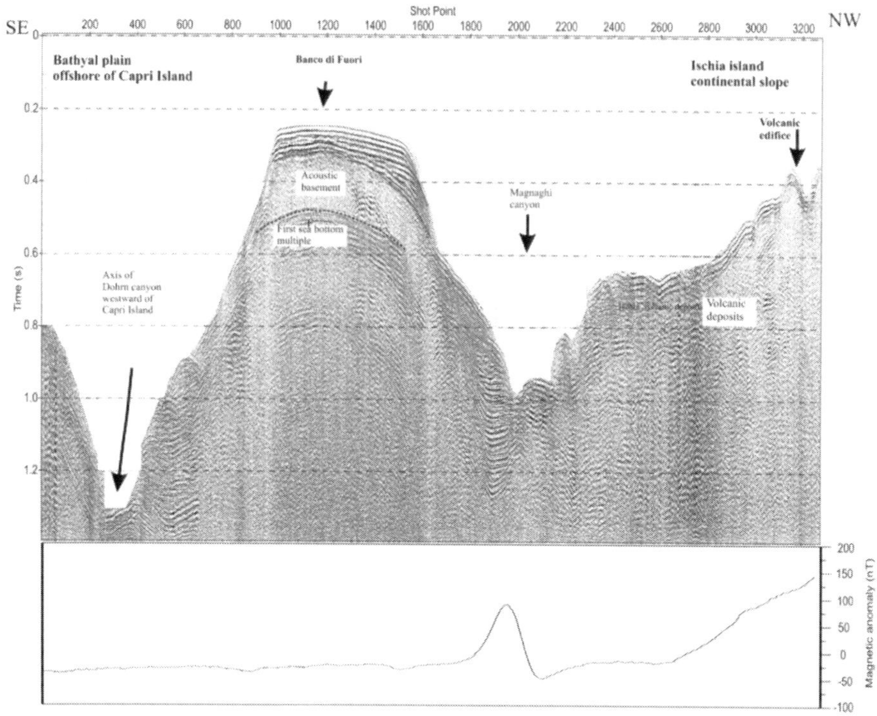

Figure 18. Seismic profile GRNA85 and corresponding interpretation. The relative magnetic profile is reported in the inset. The flat trending of the magnetic anomaly in the Banco di Fuori structural high suggests that it could represent a tilted block of Mesozoic carbonatic acoustic basement. A pronounced magnetic anomaly has been singled out on the northwestern flank of the Magnaghi canyon, incised in the volcanic deposits of the Procida island continental slope.

The seismic line GRNA85 crosses the axis of the Dohrn canyon westward of Capri island in the zone of the confluence of the two branches of the canyons, reaching water depths of more than 1100 m (Figure 18). The northern slope of the Banco di Fuori high bounds the Magnaghi canyon, reaching water depths of about 750 m. The canyon is unrelated to a present or past fluvial drainage system on land. The geological interpretation of the seismic profile GRNA85 has shown that the Magnaghi canyon incides an acoustically transparent seismic unit, interpreted as volcanic deposits, probably pertaining to the Procida volcanic complexes (55 -18 ky B.P.; Rosi and Sbrana, 1987).

The correlation between the magnetic anomaly profile and the seismic line has confirmed that the the north-western flank of the Magnaghi canyon is incised in volcanic deposits. This is shown by the trending of the magnetic anomaly along the same profile, which is flat in the first half of the line despite the complex sea floor topography, while it follows the morphology of the sea bottom, starting from the north-western flank of the Banco di Fuori high. The flat trending of the magnetic anomaly in the first part of the profile confirms the sedimentary nature of the Banco di Fuori high. The high could represent a tilted block of Meso-Cenozoic carbonates located along the structural high Sorrento Peninsula-Capri island. Moreover, the Banco di Fuori high seems to have acted as a morphological and structural barrier for the propagation of the volcanic deposits coming from the eruptive cycles of Ischia Island.

Along the northwestern flank of the Magnaghi canyon, a pronounced magnetic anomaly, in the order of 90 nT was observed. Moreover, starting from the canyon towards the Ischia continental slope the values of the magnetic anomaly increase progressively from 0 to 150 nT, confirming the volcanic nature of intensively magnetized volcanic deposits cropping out near the sea bottom. It is worth noting the correspondence between the sea floor topography and the shape of the magnetic anomaly along the Ischia continental slope.

An approach to the study of marine geophysical data based on the integrated interpretation of seismic and magnetic profiles was adopted in this study, based on the integrated interpretation of seismic and magnetic data (Aiello et al., 2005). This qualitative method improved the geological knowledge in an active volcanic area as the Gulf of Naples, since it disclosed and located several main volcanic bodies based on seismic interpretation, well constrained by the occurrence of significant magnetic anomalies (Aiello et al, 2005).

The volcanic bodies occurring in the subsurface of the Gulf of Naples are significantly linked to the eruptive phases of Somma-Vesuvius, Phlegrean Fields and Ischia and Procida islands volcanic complexes, strongly conditioning the geological setting of the Gulf of Naples. The role of the Dohrn canyon as a main morphological and structural lineament in the investigated area has also been confirmed by our data analysis. The Dohrn canyon western branch sharply separates a south-eastern sector of the Gulf of Naples, characterized by the occurrence of sedimentary units by a north-western one, where main volcanic bodies and units are localized.

In the western Gulf of Naples significant magnetic anomalies are located in a belt of submarine volcanic banks located in the external part of the Gulf of Pozzuoli, offshore the Phlegrean Fields volcanic complex (Figure 19; Aiello et al., 2005). The box 2 in Figure 19 shows as the Phlegrean Fields offshore represents a relatively complex magnetic anomaly area, characterized by several magnetic anomaly fields with different intensity. Two dipolar anomalies, characterized by a maximum-minumum couple, have been identified. The first anomaly, E-W oriented and located in the northernmost part of the area shows a minimum of – 200 nT, associated to a maximum of + 185 nT. Such a values, which are relatively not so high, could be associated with volcanic bodies not cropping out at the sea bottom, but buried by sediments. The second anomaly, NW-SE oriented and located in the easternmost part of the area shows a maximum-minimum couple with a relative intensity similar to that one of the previously discussed field. Besides these two magnetic anomaly fields, corresponding to magnetic bodies and/or volcanic edifices, other anomalies, not dipolar and of lower intensity, ranging between 40 and 135 nT are due to the occurrence of small volcanic edifices (Figure 19).

Direct sampling and age determination of the Phlegrean volcanic bank are not yet available, since the volcanic acoustic basement constituting the banks cannot be easily dredged, because they are covered by Late Pleistocene-Holocene marine sediments. Some constraints on the evolution of the Miseno and Pentapalummo volcanic banks have been already published (Pescatore et al., 1984) based on the dredging of a trachitic block in the area, correlated to the Campanian Ignimbrite pyroclastic flow deposits (Rosi and Sbrana, 1987). The identification and the activity of the Miseno and Pentapalummo volcanic banks can be probably correlated to the first phase of activity of the Phlegrean Fields volcanic complex (before than 35 ky B.P.). Moreover, the two banks seem to be older than the last glacio-eustatic regression (Wurmian regression; 18 ky B.P.), as evidenced by the occurrence of a main seismic reflector truncating the volcanic deposits of the banks and interpreted as the Wurmian erosional truncation, well correlated in the Gulf of Pozzuoli area (Pescatore et al., 1984; Aiello et al., 2012c).

It is worth noting that there is not a precise correlation between the submarine morphology of the Pentapalummo Bank and the shape of the related magnetic anomalies. It is probable that after the formation of the bank the intrusion of small volcanic bodies occurred. They seem to be magmatic bodies, elevated with respect to the top of the bank, which was terraced during the last glacio-eustatic oscillations (Milia, 1998). Several anomalies of the area are then interpreted as corresponding to small volcanic structures. At the time of their emplacement and consolidation these structures probably recorded a magnetization showing an intensity and direction different from the magnetization of the volcanic deposits of the Pentapalummo Bank.

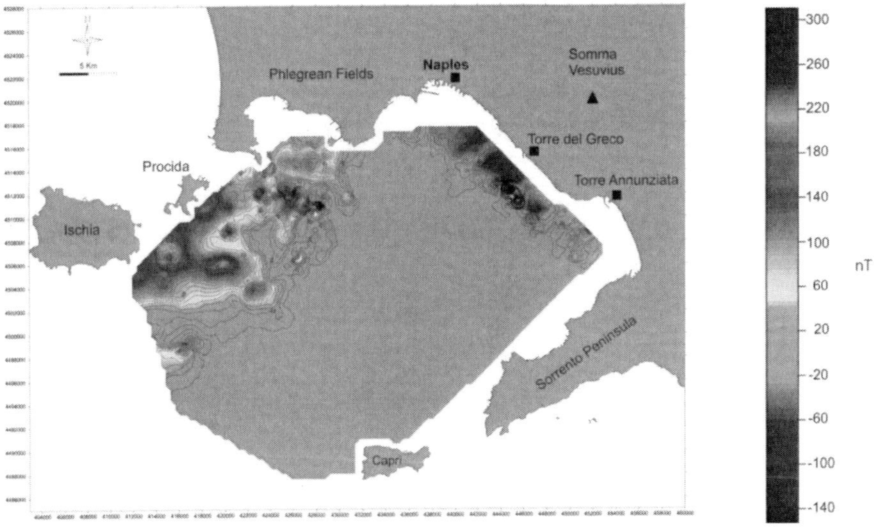

Figure 19. High resolution magnetic anomaly map of the Gulf of Naples, reporting also the main morphostructural lineaments recognized by the interpretation of Multibeam bathymetry.The box 1 represents a magnetic anomaly area located offshore the Somma-Vesuvius volcanic complex, while the box 2 represents another magnetic anomaly area located on the continental slope of the gulf, in the Magnaghi canyon's head.

A significant magnetic anomaly, in the order of 150 nT occurs in the Magnaghi canyon's head, deeply eroding the volcanic deposits of the continental slope of Procida island. This confirms that the Magnaghi canyon is entirely incised in volcanic deposits, as already evidenced by the occurrence of pronounced magnetic anomalies both in the canyon's head and its axis. On the contrary, the slope of the Gulf of Naples in the Dohrn canyon, a kilometric feature characterized by two branches joining into a main axis reaching water depths of more than 1000 m (Milia, 2000; Aiello et al., 2001) does not show magnetic anomalies, confirming that this canyon deeply erodes sedimentary units supplied by the palaeo-Sarno river mouth. An important, rounded-shaped magnetic anomaly has been identified in the continental slope of Procida island. This confirms that the Magnaghi canyon is entirely incised in volcanic deposits, as already evidenced by the occurrence of pronounced magnetic anomalies both in the canyon's heads and in its axis. On the contrary, the slope of the Gulf of Naples in the Dohrn canyon, a kilometric feature characterized by two branches joining into a main axis reaching water depths of more than 1000 m (Milia, 2000; Aiello et al., 2001) does not show magnetic anomalies, confirming that this canyon deeply erodes sedimentary units supplied by the palaeo-Sarno river mouth. An important rounded-shaped magnetic anomaly has been identified in the continental slope of the Bay, in the sector between Dohrn and Magnaghi canyons. This evidence suggests the occurrence of a previously unknown volcanic edifice, located between the two canyons. The lack of magnetic anomalies in the Banco di Fuori high, a main morpho-structural high separating the two canyons and bounding southwards the gulf, confirms that it probably represents a Mesozoic carbonatic block, located along the Sorrento Peninsula-Capri island structural elongment, resulting from the regional uplift and tilting of the acoustic basement.

5.4. South-Eastern Ischia Offshore

5.4.1. General Setting

The acquisition of a seismic grid covering the south-eastern Ischia offshore, here interpreted and discussed, was carried out by using a multielectrode Sparker system (SAM96 model; Aiello et al., 2012a). The advantages of the Multitip Sparker include shorter pulse lengths for an equivalent energy discharge, as well as an increase in peak pressure, i.e. the amplitude of the outgoing acoustic wave. The sparker source used in this survey generated 200 J in the 200-2000 frequency range. This seismic grid, covering the south-eastern sector of Ischia island facilitated the stratigraphic correlations between seismic sections and revealed structural and stratigraphic variations along the seismic lines.

The proposed stratigraphy derives from the type of data utilized in marine geology (reflection seismics) and from the methods of seismic interpretation (high resolution sequence stratigraphy). The geological structures recognized through the seismic interpretation are acoustically transparent seismic units, representing the volcanic acoustic basement and the systems tracts of the Late Quaternary depositional sequence (Fabbri et al., 2002). The depositional systems, referred respectively to the sea level fall (FST; Helland-Hansen and Gjelberg, 1994), sea level lowstand (LST) and related internal subdivisions (Posamentier et al., 1991), transgressive phase (TST; Posamentier and Allen, 1993; Trincardi et al., 1994) and sea level highstand (HST) have been identified in the Late Quaternary depositional sequence of the Ischia offshore through seismo-stratigraphic analysis.

The seismic stratigraphy is an analytical methodology for the subsurface geological survey (Vail et al., 1977; Mitchum et al., 1977). It was developed from the end of the 1970s of the last century and particularly applied to the analysis of continental margins. The seismo-stratigraphic analysis is based on the recognition of the lower and upper terminations of seismic horizons and of their geometries (onlap. erosional truncation, downlap, toplap) with respect to the main unconformities bounding specific intervals defined "depositional sequences" (Type 1 or Type 2 sequence boundaries; Vail et al., 1977). This approach offers the opportunity to reconstruct the tectono-stratigraphic evolution of a sedimentary basin, having relationships to relative sea level oscillations.

In the case of Ischia island, the occurrence of isolated volcanic bodies (intrusions, domes, volcanic necks and tabular, acoustically transparent seismic units) makes the sequence stratigraphic approach particularly complex for the geological interpretation of seismic profiles. While the volcanic bodies (i.e. lava flows, domes, intrusions) cannot be internally investigated through the reflection seismics because they are acoustically transparent, the seismic facies of pyroclastic edifices and/or buried pyroclastic deposits may be on the contrary detected, due to their internal stratification. The marine sedimentation includes both the contribution of alluvial and marine sediments and the input of volcanites and volcaniclastic deposits originating from the eruptions of Ischia and Procida volcanic complexes.

According to the interval velocities already established (D'Argenio et al., 2004; Aiello et al, 2005; 2011b; Di Fiore et al., 2011), the thickness of the Pleistocene to Holocene marine sediments in the Ischia offshore varies from 0.4 to 0.5 sec twt (corresponding to about 340-425 m) to the north of Ischia and Procida islands, to 0.2-0.3 sec twt (corresponding to about 170-255 m) in the western Ischia offshore, while the sedimentary drapes, along all the continental slope of the island, reach several tens of meters.

Based on the seismo-stratigraphic data herein discussed, the stratigraphic architecture of the Ischia-Procida marine deposits has been interpreted in terms of systems tracts of the Late Quaternary depositional sequence. The external geometry and seismic facies, together with the identification of a downlap (flooding) surface, support the interpretation of part of the seismic sequences as TST deposits. The Ischia-Procida TST sediments are characterized by fillings with onlap geometries on the flanks and on the top of the underlying volcanic units. Some of the upper intervals are instead interpreted as HST deposits, characterized by parallel to slightly prograding seismic reflectors, downlapping the underlying flooding surface.

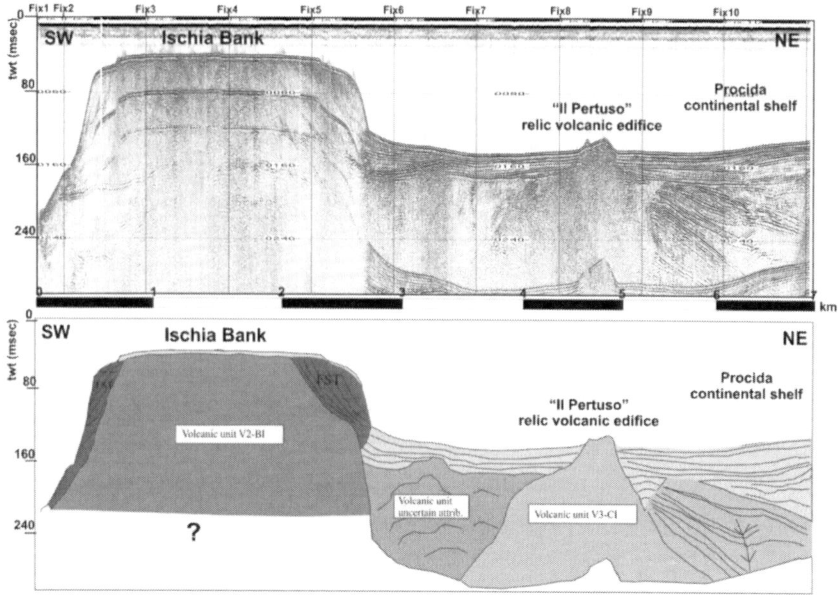

Figure 20. Seismic profile L57 located in the southeastern and eastern sectors of the Ischia island (Ischia Bank – Il Pertuso volcanic edifice – Procida continental shelf) and geologic interpretation (modified after Aiello et al., 2012a).

5.4.2. Seismic profile L57

The seismic profile L57 crosses the southeastern and the eastern sectors of Ischia. The line runs from the volcanic structure of the Ischia Bank to the Ischia Channel (Figure 20; Aiello et al., 2012a), where it crosses the "Il Pertuso" volcanic edifice and reaches the Procida continental shelf. The seismo-stratigraphic analysis has allowed the identification of volcanic and sedimentary seismic units. A volcanic acoustic basement (unit V2-BI; Figure 20; Aiello et al., 2012a), characterized by an acoustically transparent seismic facies, indicates the occurrence of lavas and pyroclastites genetically related to the main morpho-structure of the Ischia Bank. FST deposits have been identified on both the slopes of the bank (Figure 20). The top of the volcanic structure is overlain by a thin drape of bioclastic sands and gravels, cropping out at the sea bottom.

The unit of Ischia Channel (unit V3-CI; Figure 20) is made of pyroclastites and lavas genetically related to relic hydromagmatic volcanic edifices of eastern Ischia, presumably older than the Vivara volcano (Scandone et al., 1991; de Alteriis et al, 1994). Another volcanic seismic unit, probably volcanic in nature, underlies the marine deposits of the Ischia

Channel (volcanic unit, uncertain attribution; Figure 20; Aiello et al., 2012a). In correspondence to the "Il Pertuso" volcanic structure, this unit appears in lateral contact with the volcanic sequence of the Ischia Channel (unit V3-CI; Figure 20; Aiello et al., 2012a).

Proceeding towards the Procida shelf, the V3-CI volcanic sequence grades laterally into a relic wedge, probably Pleistocene in age, characterized by prograding clinoforms (Figure 20). This wedge is overlain by a pyroclastic unit, onlapping depressions and channel-type erosional morphologies, genetically related to the last eruptive phases of the Procida Island.

5.4.3. Seismic profile L50

The seismic profile L50 crosses the southwestern offshore of the Ischia island, from the continental shelf, west of Punta Imperatore, to the continental slope located southward of Capo Negro. The line meets three main submarine canyons (Figure 21; Aiello et al., 2012a), pertaining to the southern Ischia canyon system, respectively located at water depths of 173 m (canyon 1), 240 m (canyon 2) and 285 m (canyon 3). The three canyon thalwegs are bounded by levees that may evolve into extensive channel-levee complexes. A volcanic acoustic basement has been identified (Figure 21; Aiello et al., 2012a), dipping from the continental shelf off Punta Imperatore towards the southwestern Ischia offshore. Based on land-sea correlation, this basement could be linked to the lavas cropping out at Punta Imperatore (southwestern Ischia), represented by alkali-trachytic grey lavas, dating to about 116-123+/- 6 ky (Vezzoli 1988). The top of the volcanic acoustic basement appears strongly incised, probably in a subaerial environment. A wide pyroclastic edifice has been recognized, buried in the continental shelf (Figure 21; Aiello et al., 2012a). This edifice is interstratified in the FST deposits, recognized on the continental shelf off Punta Imperatore, where Quaternary sediments are truncated by an erosional unconformity, overlain by thin highstand deposits. A relic Pleistocene prograding wedge has also been identified in correspondence to the western canyon, offshore southwestern Ischia, where it overlies the volcanic acoustic basement corresponding to the Punta Imperatore lavas. This wedge is overlain by a marine seismic unit, which seems to be syntectonic, since it appears to be deposited during the activity of the normal fault downthrowing the volcanic acoustic basement.

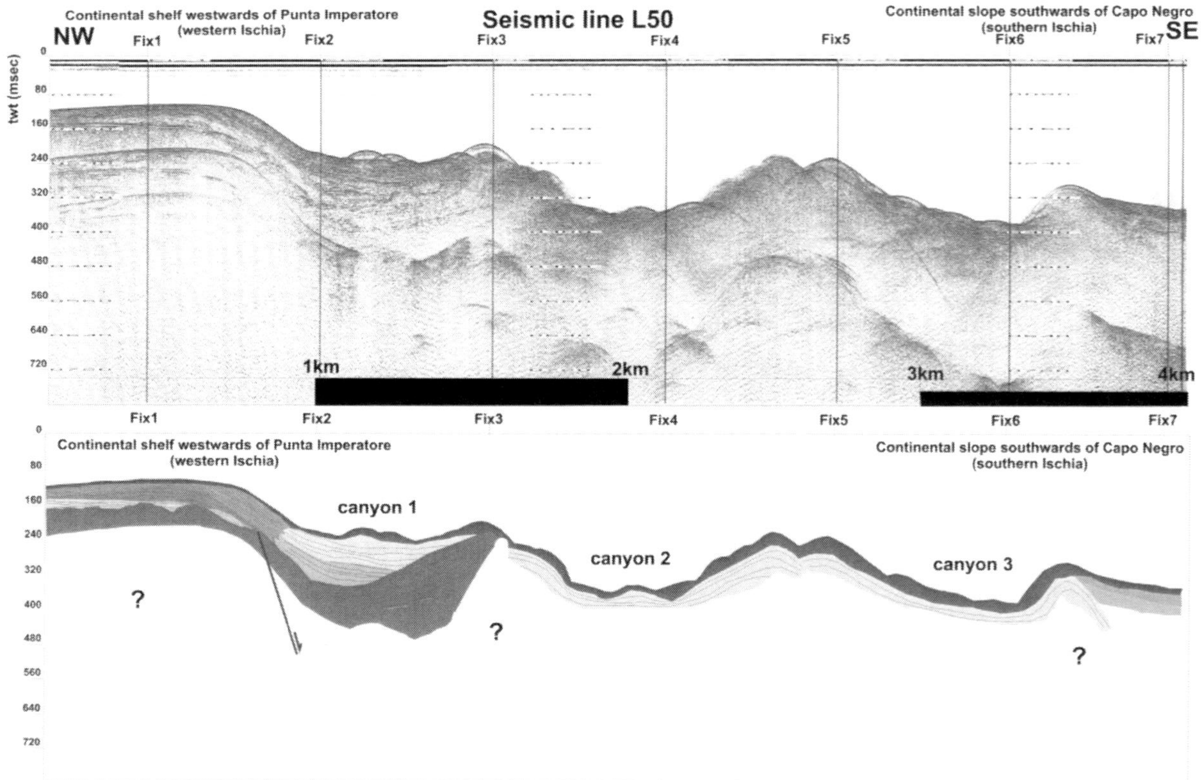

Figure 21. Seismic profile L50, located in the southwestern sector of Ischia island and corresponding geologic interpretation (modified after Aiello et al., 2012a). Note the occurrence of three main canyons, pertaining to the southern Ischia canyon system.

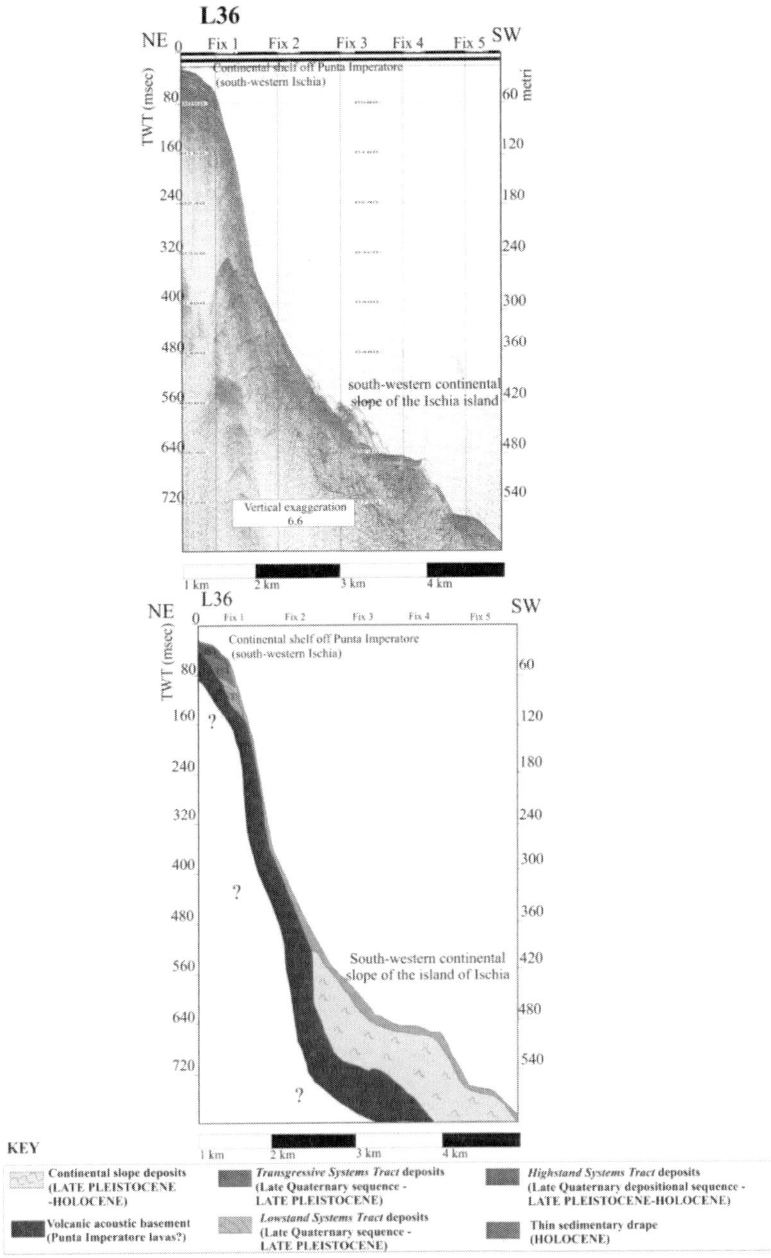

Figure 22. Seismic profile L36, located in the southwestern sector of Ischia island and its geologic interpretation. The profile runs from west to east in the southwestern Ischia offshore. Depth conversion on the right applies to the sea bottom.

5.4.4. Seismic profile L36

The seismic profile L36 runs from the continental shelf of Punta Imperatore (south-western Ischia) towards the continental slope (Tyrrhenian sea), where an inclined volcanic acoustic basement has been identified. A thin sedimentary drape covers the acoustic basement in the whole continental slope, where a thick sedimentary unit has been recognized at the toe of the slope, probably related to slope deposits. On the continental shelf, HST, TST and LST deposits have been identified (Figure 22).

5.4.5. Seismic profile L38

The seismic profile L38 crosses the southwestern Ischia offshore (Forio d'Ischia) towards the southwestern Ischia slope. In particular, the section has been recorded on the continental shelf off Grotta del Mavone (Figure 23), one of the largest cavities of Ischia (D'Ambra 1992). This cavern is located in the homonymous bay, next to the Campotese wood (Forio D'Ischia; Figure 23). Off the Grotta del Mavone, the continental shelf is narrow and the depositional shelf break is located at 35-40 m (Figure 24). The continental slope is steep and shows hints of mass wasting. The seismo-stratigraphic analysis allowed to identify a volcanic acoustic basement under the continental slope, genetically related to the Grotta del Mavone lava domes and adjacent volcanic rocks (Figure 24). The stratigraphic architecture of the continental shelf is characterized by the Late Quaternary depositional sequence (progradational HST, TST and LST deposits).

Figure 23. The southern Ischia coastal cliff in correspondence to the Grotta del Mavone cave (modified after Aiello et al., 2012a).

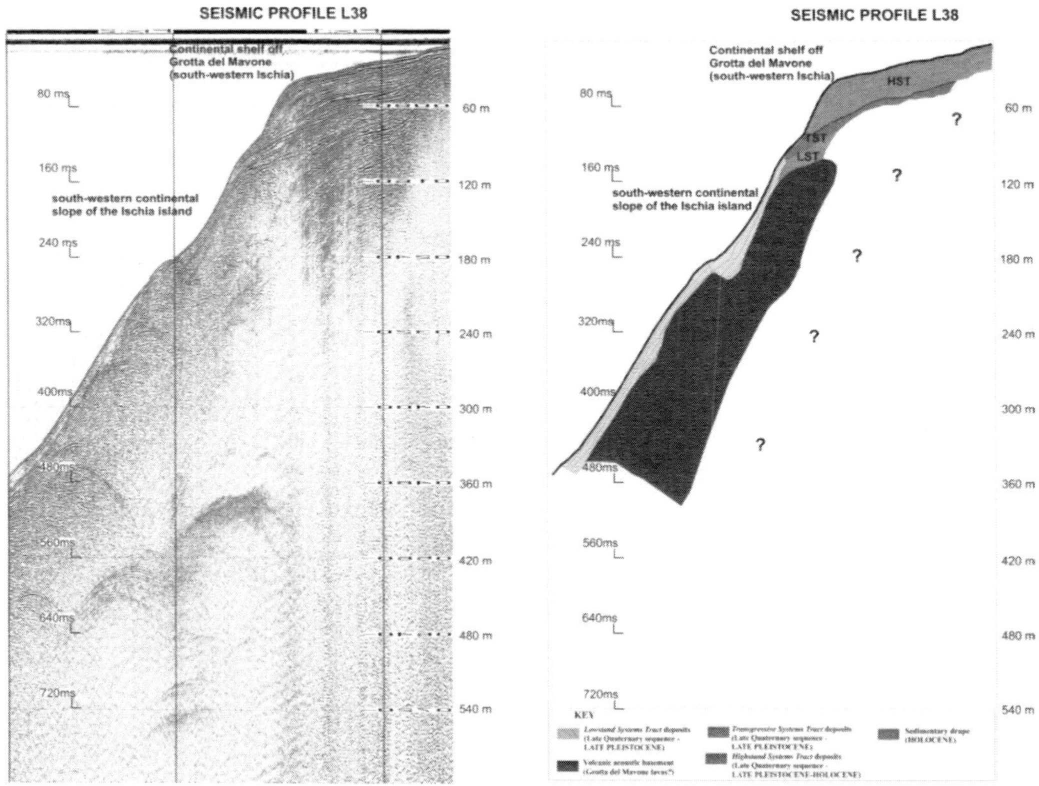

Figure 24. Seismic profile L38 located in the southwestern sector of Ischia island, offshore the Grotta del Mavone cavity and its geologic interpretation.

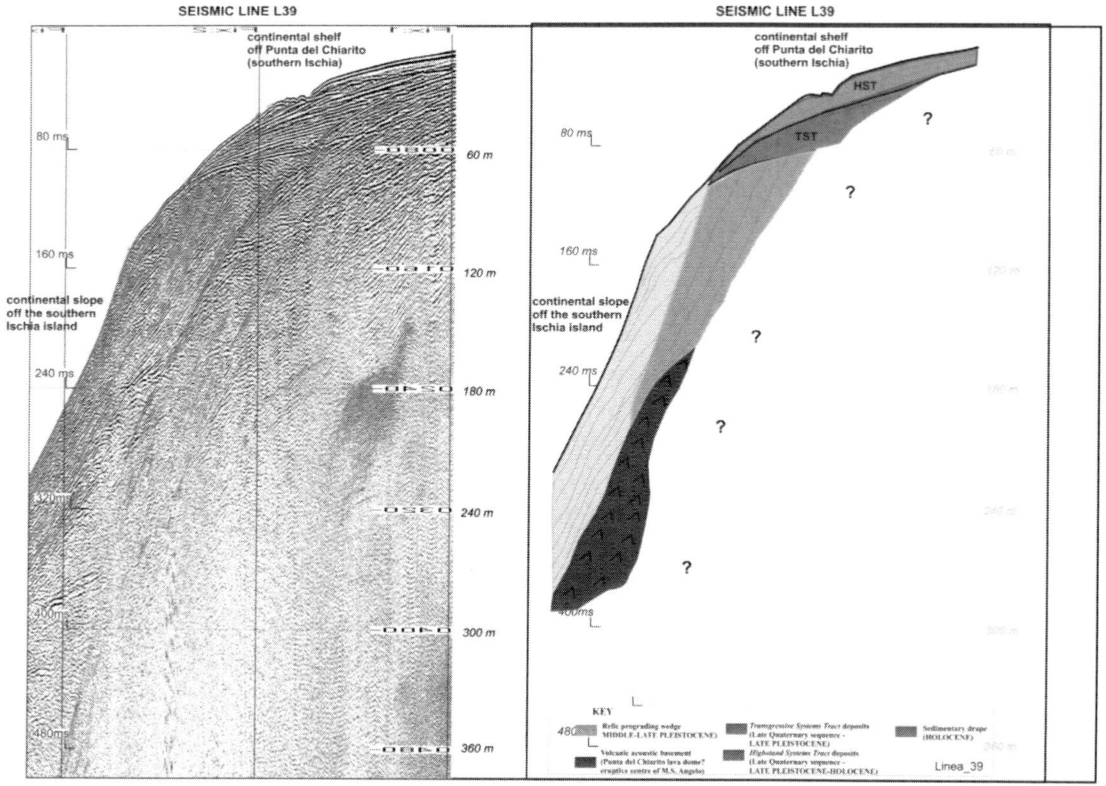

Figure 25. Seismic profile L39 recorded in the Southern Ischia offshore (Punta del Chiarito) and corresponding geologic interpretation (modified after Aiello et al., 2012a).

5.4.6. Seismic profile L39

The seismic profile L39 crosses the southern Ischia offshore (Punta del Chiarito) towards the southern Ischia slope. A volcanic sequence of well-stratified deposits crops out between Punta del Chiarito and Punta S. Angelo (Vezzoli, 1988). It is formed of tuffs and pumiceous lapilli as well as pomiceous tuffs and breccias, originating from pyroclastic fluxes and surges and attributed to the Faro di Punta Imperatore Formation. Dome-shaped components of the volcanic basement crop out at the sea bottom of the inner shelf at about 30 m bsl (Budillon et al. 2003b). At Punta del Chiarito and Capo Negro, the products of explosive eruptions are associated with the lavas. These products are overlain by lava accumulations that, being resilient to the volcanic eruptions more than other materials, have originated the three promontories (Punta del Chiarito, Capo Negro and Punta San Pancrazio).

The seismo-stratigraphic analysis of the L39 seismic section (Figure 25; Aiello et al., 2012a) has revealed the occurrence on the continental slope of the volcanic acoustic basement, genetically related to the Punta del Chiarito lava dome (eruptive center of Monte Sant'Angelo). This substratum is unconformably overlain by a wide relic prograding wedge, probably Middle-Late Pleistocene in age. On the slope a thick wedge of Holocene deposits unconformably overlies the relic progradational unit. On the continental shelf, the TST deposits are characterized by retrogradational patterns and are unconformably overlain by HST deposits, with progradational geometries.

5.4.7. Seismic profiles L41, L42 and L43

Three seismic profiles have been recorded off the Maronti littoral belt and toward the continental slope, which in this sector appear steeper than in the adjacent offshore areas of southern Ischia (Figure 26; Aiello et al., 2012a). The seismic profile L41, recorded in the western Maronti sector, shows in continuity the volcanic acoustic basement from the shelf to the slope (Figure 26a). A lava dome, buried below 300-360 m of volcaniclastic deposits, has been identified. TST and HST deposits characterize the stratigraphic architecture of the narrow continental shelf. The steep slope is draped by a Holocene wedge thickening from 180 to 360 m of water depth. The seismic line L42 has been recorded in the center of the bay (Figure 26b) and similar seismo-stratigraphic units have been recognized on the L42 line. The volcanic acoustic substratum crops out along the slope under a thin Holocenic drape, while a relevant canyon's head has been recognized on the upper slope. The

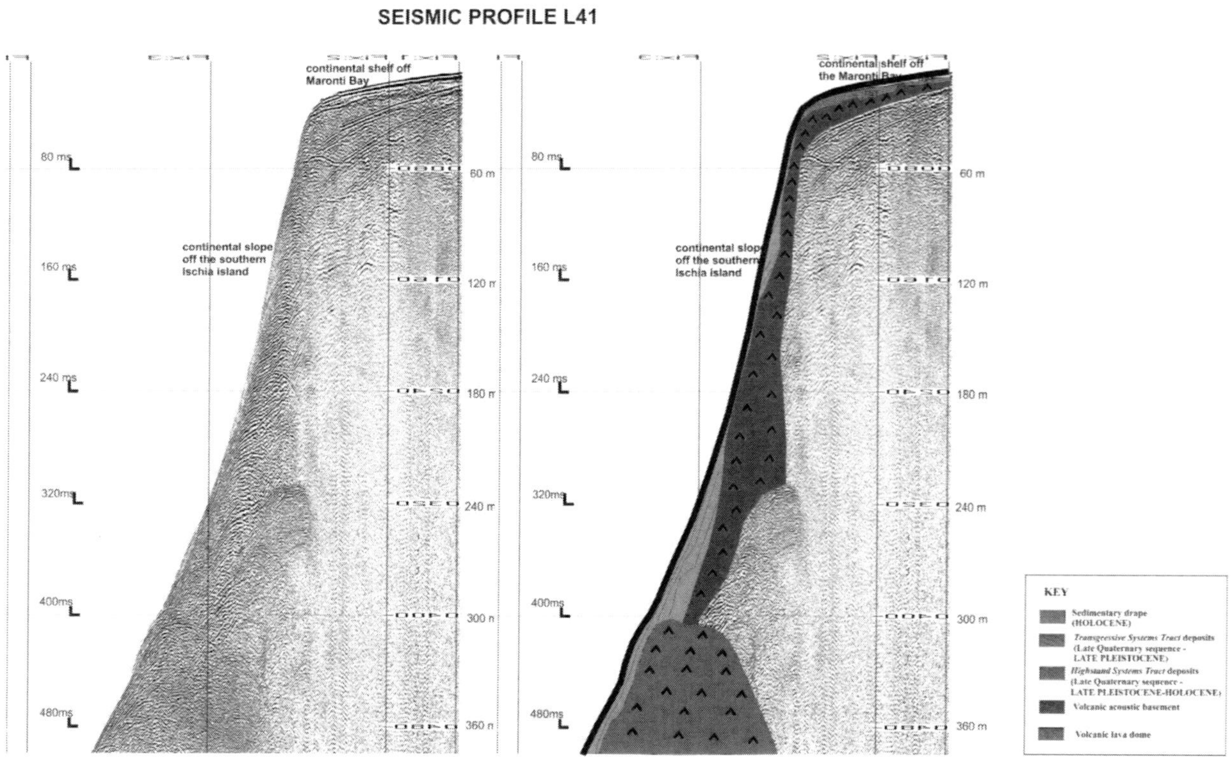

Figure 26a. Seismic profile L41 and corresponding geologic interpretation (modified after Aiello et al., 2012a).

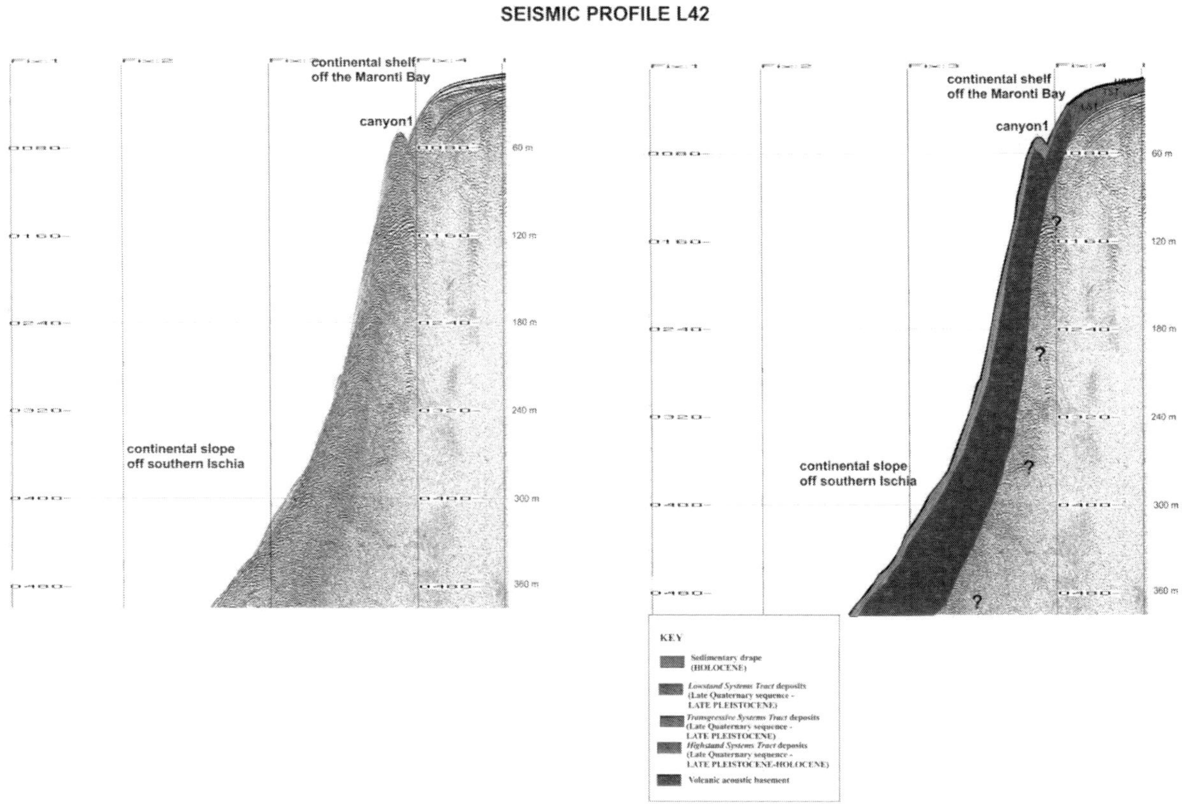

Figure 26b. Seismic profile L42 and corresponding geologic interpretation (modified after Aiello et al., 2012a.

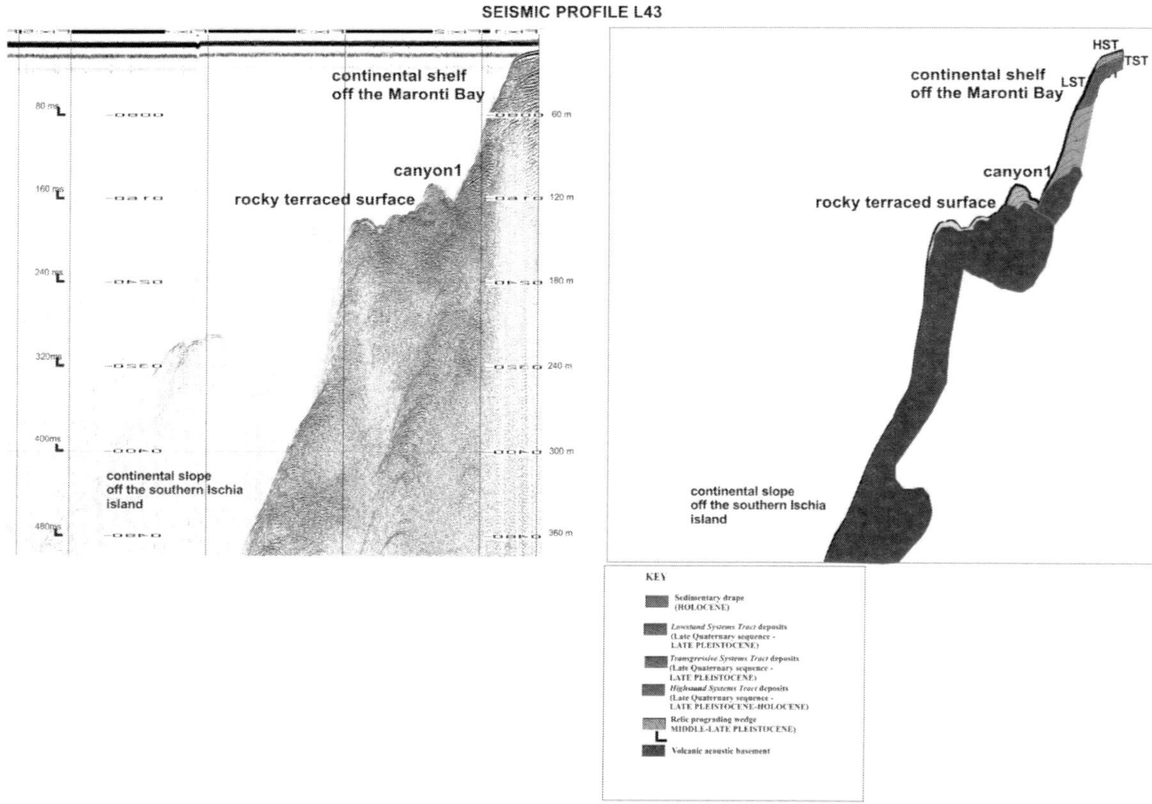

Figure 26c. Seismic profileL43 recorded off the Maronti coastal area and its geologic interpretation (modified after Aiello et al., 2012a).

seismic profile L43 has been recorded in the eastern Maronti sector (Figure 26c). The stratigraphic architecture of the narrow shelf is characterized by LST, TST and HST deposits. On the upper slope, a rocky terraced surface has also been recognized, at water depths ranging from 140 to 180 m, while a canyon head has been recognized at the foot of the slope. Also in this case, on the narrow continental shelf, LST, TST and HST deposits have been identified.

5.4.8. Seismic profiles L44, L45 and L46

The seismic profiles L44, L45 and L46 have been recorded in the Barano Bay in correspondence to the Scarrupata di Barano (Figure 27; Aiello et al., 2012a). This area has been recently studied in detail through geophysical techniques, which have been applied to the landslide of Monte di Vezzi (Di Maio et al., 2007) that occurred in April 2006 involving the northern slope of Ischia (Di Nocera et al., 2007). The Monte di Vezzi belongs to a ridge stretching in a counter-Apenninic (NE-SW) direction, bordered to the southeast by a regional normal fault that also originated in the Ischia graben (Vezzoli, 1988). The succession of volcanic events is represented along the outcrop of the Scarrupata di Barano marine cliff (Vezzoli 1988; Di Maio et al. 2007). In the stratigraphic section, domes and lava flows of the pre-Green Tuff activity crop out (La Guardiola lavas, upper and lower Scarrupata di Barano Formation, domes and lava flows of Monte di Vezzi). During the eruptive phases of the post-Green Tuff cycles, these lavas have been covered by the products derived from several pyroclastic eruptions of other centers, as follows: (a) Monte di Vezzi – not welded pumice flows from the Monte Epomeo Green Tuff; (b) S. Costanzo Formation, made up of well-stratified, whitish tuffs with pumices and lapilli; (c) Monte di Vezzi Formation, consisting of bedded pumice fall breccias and brown scoriaceous layers and (d) Piano Liguori Formation, made of interbedded white ashes and pumice layers. The seismic profile L44 (Figure 27a) has been recorded in the western Barano sector. Here a narrow continental shelf, characterized by the Late Quaternary depositional sequence (LST, TST and HST deposits) grades into the Scarrupata di Barano, ranging in depth between 100 and 540 m. The volcanic acoustic substratum is very thick and appears organized into two main seismic units (Barano 1 and Barano 2 seismic units) separated by an erosional unconformity. They should be genetically related with the upper and lower Scarrupata di Barano formations, both composed of tuffs and tuff-breccias. A thick seismic sequence, characterized by channel-levee complexes and deposited at the toe of the Scarrupata di Barano, has been also recognized at water depths between 300 and 540 m. Here, the upper slope is characterized by a thin Holocene drape, overlying the volcanic substratum of the Barano 1 unit.

The seismic profile L45 (Figure 27b), recorded in the central Barano sector, shows a volcanic substratum, organized into the already mentioned, acoustically transparent seismic sequences Barano 1 and Barano 2. A submarine canyon incises the volcanic deposits of the Barano 1 sequence at water depths of 360 m. In correspondence with the canyon, a normal fault downthrows the volcanic deposits.

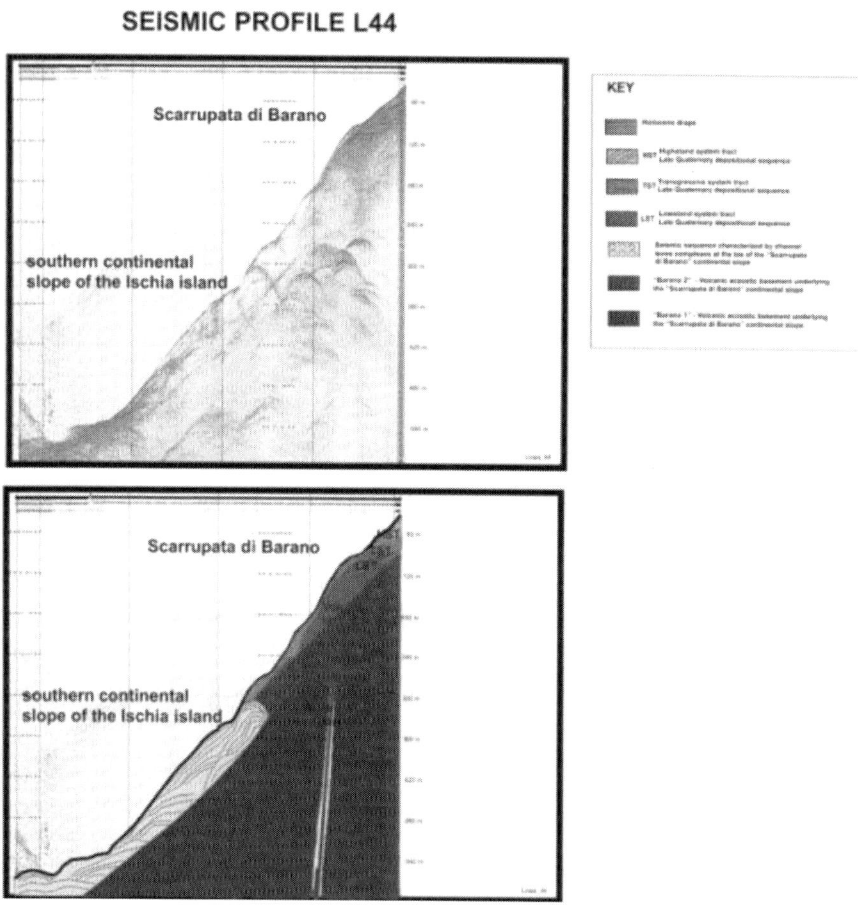

Figure 27a. Seismic profile L44 and corresponding geologic interpretation (modified after Aiello et al., 2012a).

Downslope, at water depths ranging between 360 and 440 m, a thick seismic sequence deposited at the foot of the Scarrupata di Barano and characterized by channel-levee complexes has been observed. Finally, the seismic profile L46, recorded in the eastern Barano sector, has shown a buried pyroclastic volcanic edifice, interstratified with the highstand sediments, while the transgressive deposits correspond to a thin retrogradational sequence. The lowstand deposits are characterized by prograding reflectors, erosionally truncated at their top. In the same Barano sector of the continental slope, a thick turbidite sequence occurs, showing parallel reflectors with an interstratified landslide, evidenced by disrupted and chaotic seismic reflectors.

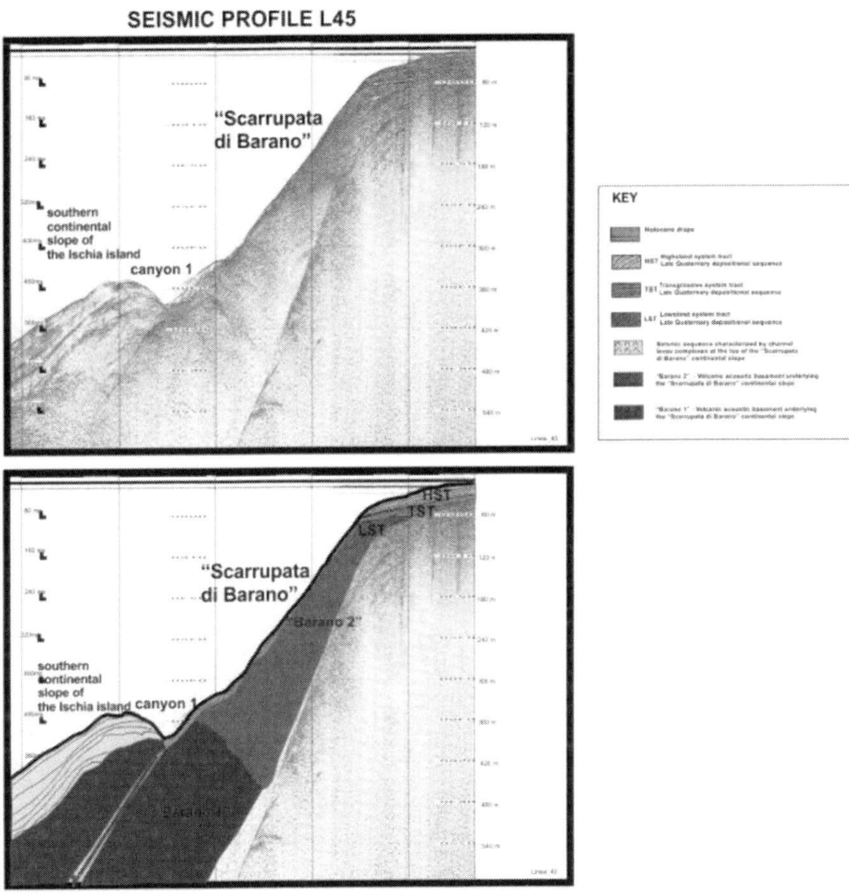

Figure 27b. (Continued)

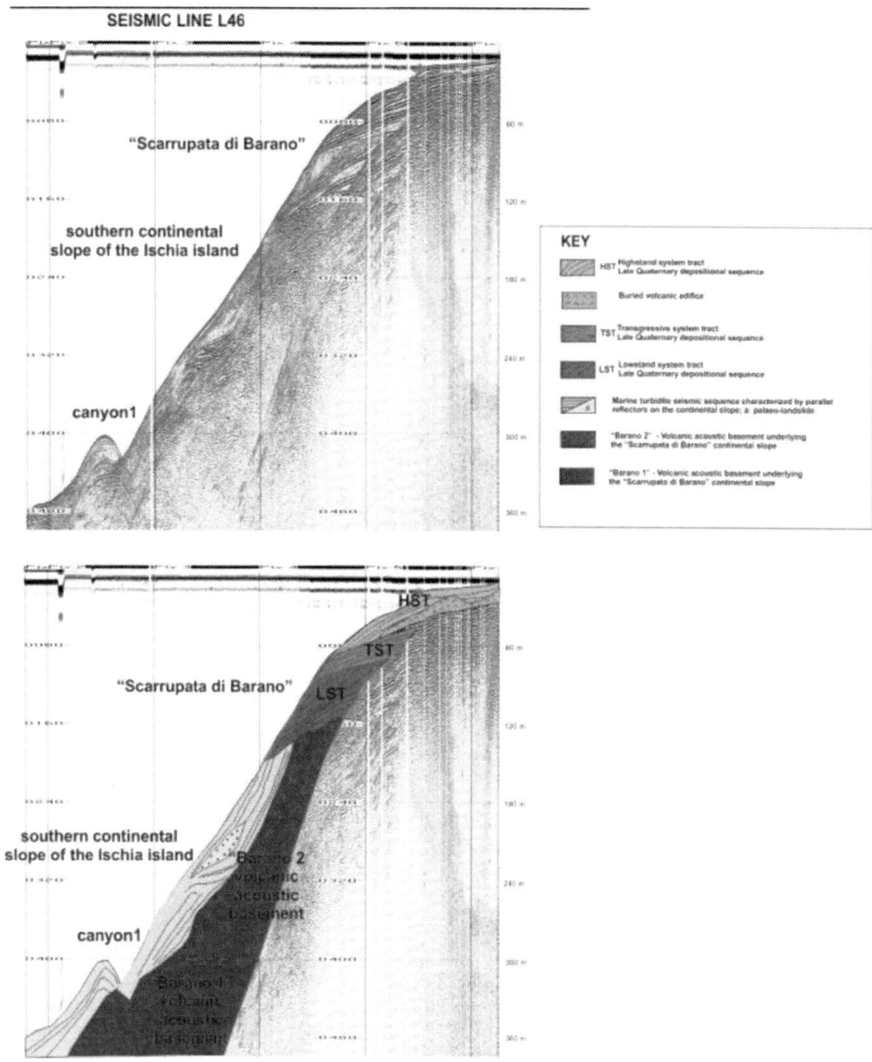

Figure 27c. Seismic profiles L45 (b) and L46 (c) recorded off the Scarrupata di Barano and its geologic interpretation (modified after Aiello et al., 2012a).

5.4.9. Seismic profile L47

The seismic profile L47 has been recorded from the Punta San Pancrazio offshore toward the Tyrrhenian sea (southeastern Ischia continental slope) at water depths ranging between 50 and 360 m (Figure 28; Aiello et al., 2012a). The oldest products cropping out onshore testify that an explosive activity

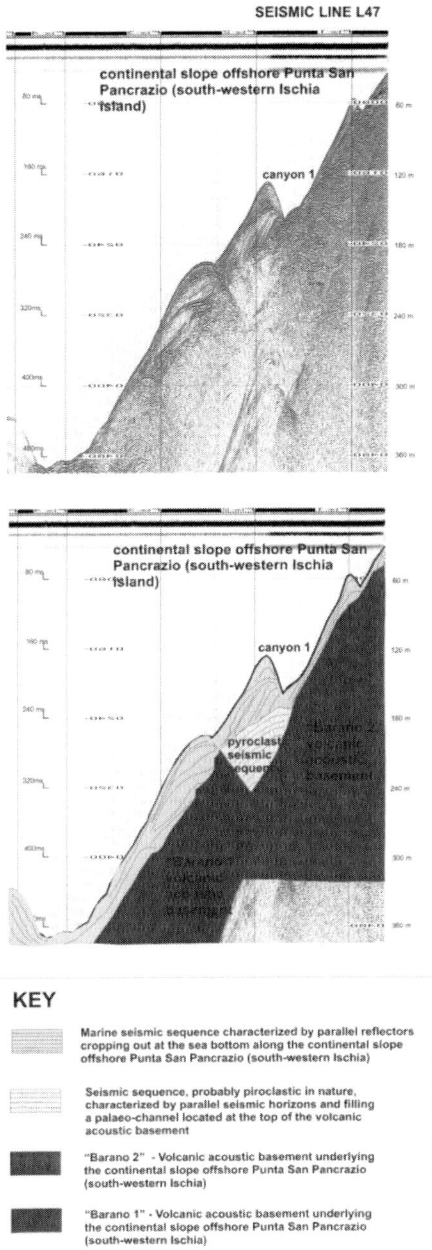

Figure 28. Seismic profile L47 recorded in the southeastern Ischia offshore at Punta San Pancrazio and corresponding geologic interpretation (modified after Aiello et al., 2012a).

occurred in correspondence to this corner of the island. They are made of pumices and scorias ("Formazione di Carrozza"), forming the internal structure of Monte di Vezzi and crop out along the slope toward the San Pancrazio beach.

Three main explosive phases followed, whose products represent the lower interval of the Scarrupata di Barano. They are represented by about 100 m of pyroclastic flow deposits, with ashes including pumices. Seismic stratigraphy features are similar to those observed in the Barano area. The volcanic acoustic basement is organized into two main seismic units (Barano 1 and Barano 2), separated by an erosional unconformity.

A wide palaeo-canyon has been rccognized at the top of the volcanic units. It is filled with pyroclastic deposits, characterized by acoustic facies from transparent to chaotic. A marine turbidite seismic sequence, characterized by parallel to progradational reflectors, crops out at the sea bottom- The San Pancrazio canyon has been clearly recognized on the seismic line. The thalweg of the canyon reaches a present water depth of about 150 m. On the flank of the canyon, a leveed seismic sequence develops.

5.4.10. Seismic profile L51

The seismic profile L51 has been recorded in the eastern Ischia offshore, starting from the Tyrrhenian slope toward the Magnaghi canyon (Figure 29; Aiello et al., 2012a). The line continues in the Ischia Channel, crossing "I Ruommoli" volcanic edifice and terminates in correspondence of the Aragonese Castle (Ischia Ponte). The Ischia Channel separates the Ischia Island from the Vivara inlet and represents the narrowest entry into Naples Bay, linking the latter to the Gaeta Gulf. The channel extends in a NW-SE direction with a minimum width of about 2.35 km between the inlet of Aragonese Castle and Punta d'Alaca, the western margin of the Vivara inlet. The sea bottoms have generally a low water depth, reaching a maximum at about 14-15 m in the narrowest point between the two islands. The so-called "Formiche di Vivara" becomes visible: a group of rocky inlets or non-emerging rocky outgrowths on the sea bottom, representing the uplifted part of a saddle reaching 4-5 m of water depth. A volcanic substratum, of uncertain attribution, underlies the volcanic unit of "I Ruommoli", formed by two mound-shaped coalescent volcanic edifices (Figure 13). The corresponding seismic unit, acoustically transparent, is herein called RM volcanic unit. It is unconformably overlain by the V3-CI unit (volcanic unit of the Ischia Channel) whose top is eroded by a subaerial unconformity, giving rise to small intra-platform basins, infilled by the TST deposits. On the eastern continental shelf of Ischia, from the Aragonese Castle to the Ischia Channel, highstand deposits appear to be well developed and range in average thickness between 40 and 20 m.

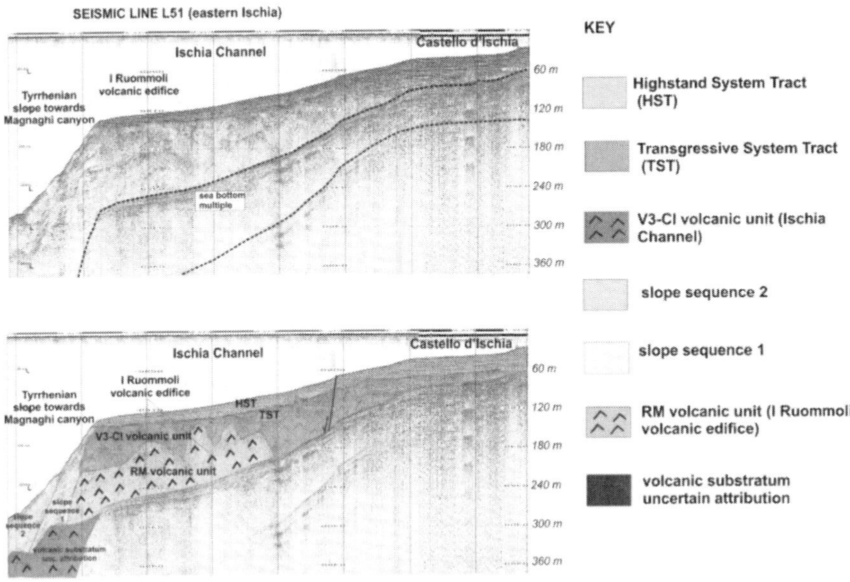

Figure 29. Seismic profile L51 recorded between the Ischia Channel and the Procida continental shelf and its geologic interpretation (modified after Aiello et al., 2012a).

5.5. North-Western Ischia Offshore

5.5.1. Seismic profile L27

In the northern offshore of Ischia, a simplified stratigraphic scheme based on the interpretation of a high resolution seismic profile has been drawn up (L27; Aiello et al., 2009a; Figure 30). It is briefly discussed here to clarify our geological understanding of the Casamicciola hummocky deposits and related debris avalanche.

The L27 line runs from the continental shelf, off Casamicciola (northern Ischia), toward the northern regression prograding wedges (FST), pertaining to the Late Quaternary depositional sequence, have been identified on the continental shelf. Debris avalanche deposits (Casamicciola debris avalanche; Budillon et al., 2003a; Aiello et al., 2009a; de Alteriis and Violante 2009), having a wedge-shaped external geometry and chaotic facies, are arranged into two distinct, superimposed bodies (H1 and H2; Figure 30). The hummocky deposits formed during two main, distinct, volcano-tectonic events occurring on the continental shelf off Casamicciola. The two bodies, H1 and H2 are characterized by facies hetheropy with the upper seismic unit of the basin filling (unit E; Figure 30).

The lower seismic unit of the basin filling is characterized by reflectors having a parallel seismic facies and shows bidirectional onlaps in correspondence to depressions formed at the top of the underlying seismic unit (unit C; Figure 30).

The intermediate seismic unit of the Ischia northern offshore (unit D; Figure 30) is characterized by parallel-to-subparallel seismic reflectors. This unit shows a strong wedging in correspondence to a normal fault (fossilized by the erosional unconformity located at the top of the unit) and facies hetheropy with the upper part of dome-shaped, buried volcanic structures (unit V2; Figure 30).

The upper seismic unit of the basin filling, off northern Ischia island (unit E; Figure 30) is characterized by parallel to subparallel seismic reflectors and locally, by prograding clinoforms. The unit is slightly downthrown by a normal fault and shows facies hetheropy with the lower part of dome-shaped buried volcanic stuctures (unit V2; Figure 30). In addition, even if the vertical displacement of the unit E by the normal fault is negligible, the occurrence of the fault is suggested by both the seismic facies (since the seismic reflectors are disrupted) and by the strong thickness variation of the unit E towards the northern Ischia offshore (Figure 30). Moreover, based on seismo-stratigraphic evidence, the suggested fault line represents a preferential way for the emplacement of the volcanic domes of Casamicciola (unit V2; Figure 30). These dome-shaped volcanic edifices are interstratified in both the lower and the intermediate seismic units of the basin filling.

Dome-shaped volcanic edifices are in lateral contact with the lower seismic unit of the basin filling (unit C) and partly with a second unit (unit D). The volcanic bodies are saturated by the erosional truncation at the top of the unit D (Figure 30). An undetermined volcanic unit off Casamicciola, corresponding to the acoustic substratum (unit V1; Figure 30) has also been identified. The volcanic acoustic basement shows seismic facies hetheropy with the unit C of the basin filling (Figure 30) and is eroded at the top by a subaerial unconformity.

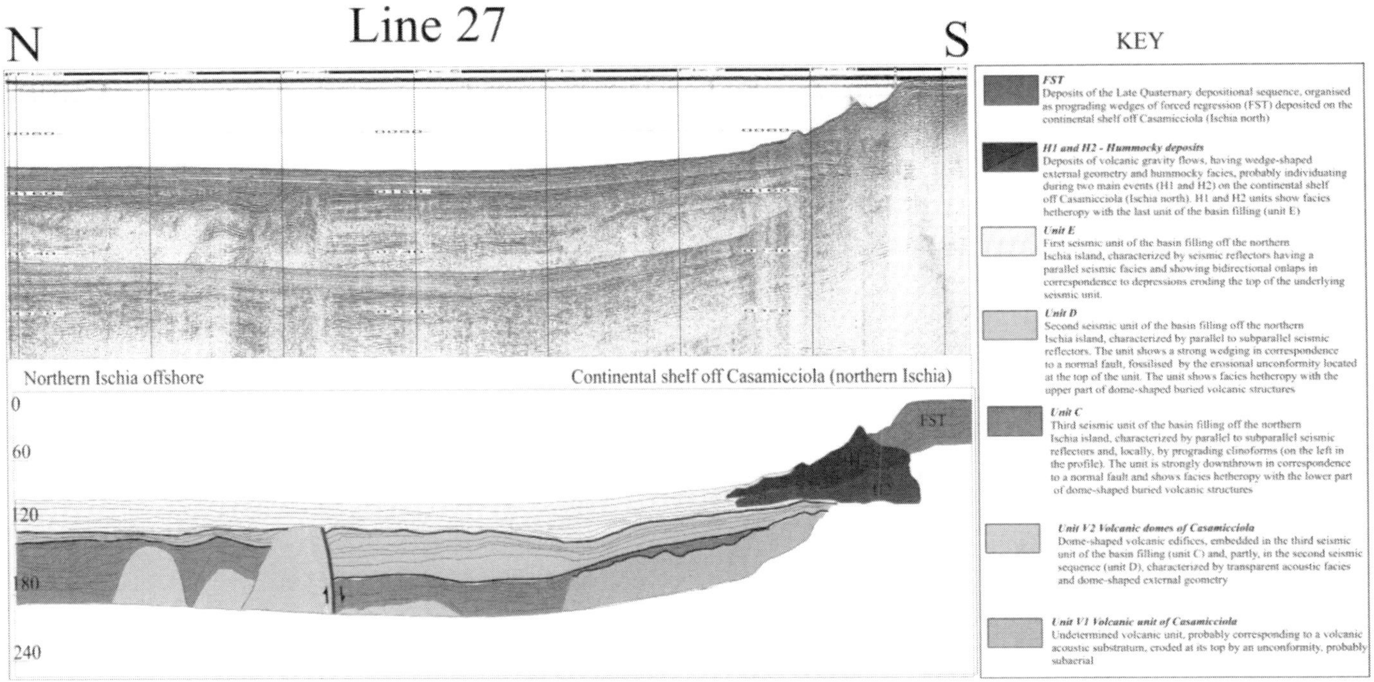

Figure 30. Seismic profile L27 recorded offshore northern Ischia by Sparker Multitip seismic source and corresponding geologic interpretation (modified after Aiello et al., 2009; 2012a). The line runs from the continental shelf off Casamicciola (northern Ischia) toward the northern offshore Ischia. Forced regression prograding wedges characterize the stratigraphic architecture of the continental shelf (FST). Two superimposed wedges of chaotic deposits suggest a multi-phase event in the evolution of the debris avalanche off Casamicciola (H1 and H2). Note that three seismic units, separated by regional unconformities, occur in the basin filling.

5.5.2. Seismic profile L28

The seismic profile L28 has been recorded offshore northern Ischia by a Sparker Multitip seismic source. The line runs from the continental shelf off Casamicciola (northern Ischia) towards the northern Ischia offshore. A qualitative sketched stratigraphic diagram of northern offshore Ischia has been constructed based on the geological interpretation of the L28 seismic profile (Figure 31).

Several seismic units have been distinguished based on seismic interpretation. The volcanic unit of Casamicciola is an undetermined volcanic unit, probably corresponding to a volcanic acoustic substratum, eroded at its top by an unconformity, probably subaerial. The volcanic domes of Casamicciola are represented by dome-shaped volcanic edifices, characterized by an acoustically-transparent seismic facies and a dome-shaped external geometry, underlying the lower seismic unit of the basin filling (unit 3 in Figure 31). The lower seismic unit of the basin filling is characterized by parallel to sub-parallel seismic reflectors. The upper seismic unit of the basin filling is also distinguished by parallel to sub-parallel seismic reflectors. The hummocky deposits (H1, H2) have also been interpreted, characterized by a wedge-shaped external geometry and a chaotic facies, probably individuated during two main volcano-tectonic events occurring on the northern Ischia offshore off Casamicciola.

The hummocky area off northern Ischia is evident on shaded relief maps of Multibeam bathymetry. The hummocky topographies are located at water depths ranging from 20 m to 180 m and show large blocks, having average dimensions in the order of hundred of meters, cropping out at the sea bottom and partly covered by Holocene sediments. The deposits form a large field having a sub-circular shape in plan view, which is bounded by depositional levees; a network of erosional gullies develops in the portion of the sea bottom occupied by the hummocks. As suggested by the interpretation of seismic profiles, the hummocky deposits (H1 and H2 in Figure 31) are interpreted as the result of two events of debris avalanches, partly superimposed.

The sketch stratigraphic scheme of Figure 31 has shown the stratigraphy of the northern Ischia offshore. The debris avalanche deposits, having a wedge-shaped external geometry and chaotic facies, are arranged in two distinct, superimposed bodies (H1 and H2 in Figure 31). The two wedges overlie the upper seismic unit of the basin filling. The lower seismic unit of the basin filling (Figure 31) is characterized by reflectors having a parallel seismic facies. The upper one shows similar seismic characteristics. Dome-shaped volcanic units and undetermined volcanic units have also been recognized.

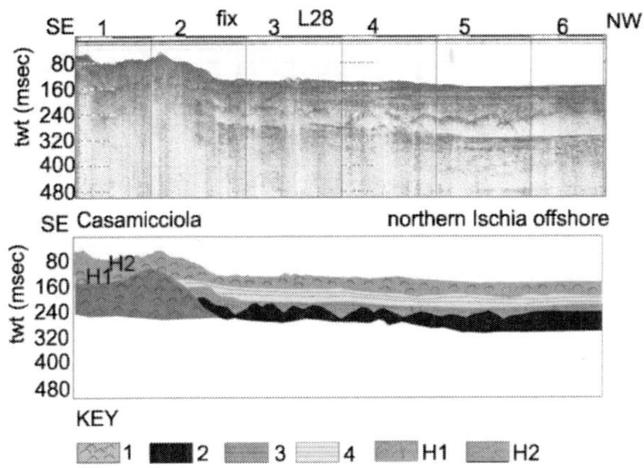

Figure 31. Seismic profile L28 recorded from Casamicciola to northern Ischia offshore and corresponding geologic interpretation. Key. 1. Volcanic unit of Casamicciola. Undetermined volcanic unit, probably corresponding to a volcanic acoustic substratum, eroded at its top by an unconformity, probably subaerial. 2. Volcanic domes of Casamicciola. Dome-shaped volcanic edifices, characterized by an acoustically-transparent seismic facies and dome-shaped external geometry, underlying the lower seismic unit of the basin filling (unit 3). 3. Lower seismic unit of the basin filling, characterized by parallel to sub-parallel seismic reflectors. 4. Upper seismic unit of the basin filling, characterized by parallel to sub-parallel seismic reflectors. H1, H2: Hummocky deposits, characterized by wedge-shaped external geometry and chaotic facies, probably individuated during two main volcano-tectonic events occurring on the northern Ischia offshore off Casamicciola.

5.5.3. Seismic profile L33

The seismic profile L33 was recorded from the eastern Ischia offshore in correspondence to Forio d'Ischia towards the Tyrrhenian offshore (Figure 32). A sketch stratigraphic scheme of the eastern Ischia offshore has been constructed based on seismic interpretation, aimed at showing the stratigraphy of the eastern Ischia hummocky deposits interpreted as debris avalanches and related volcanic acoustic substratum and marine seismic sequences.

An undetermined volcanic substratum has been recognized, overlain on the continental shelf offshore Forio d'Ischia, by a chaotic seismic sequence interpreted as hummocky deposits offshore Forio d'Ischia. In the surrounding basin several marine seismic sequences have been recognized through the geologic interpretation of seismic profile. At the shelf margin the upper seismic sequence is interstratified to forced regression/lowstand deposits, showing the stratigraphic architecture of prograding wedges.

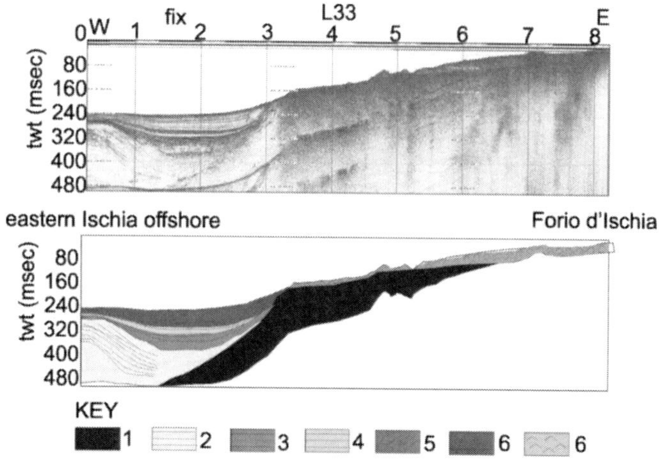

Figure 32. Seismic profile L33 recorded from Forio d'Ischia to western Ischia offshore and corresponding geologic interpretation. Key. 1. Volcanic unit of Forio d'Ischia. Undetermined volcanic unit, probably corresponding to a volcanic acoustic substratum, eroded at its top by an unconformity, probably subaerial. 2: Lower seismic unit of the basin filling, characterized by aggradational seismic reflcctors. 3. Intermediate first seismic unit of the basin filling, characterized by parallel to sub-parallel seismic reflectors. 4. Intermediate second seismic unit of the basin filling, characterized by parallel to sub-parallel seismic reflectors. 5: Forced regression/lowstand prograding wedges. 6: H1: Hummocky deposits, characterized by a wedge-shaped external geometry and chaotic facies, individuated during a main volcano-tectonic event on the western Ischia offshore off Forio d'Ischia.

Several seismic units have been recognized based on the geologic interpretation of the seismic profile. The volcanic unit of Forio d'Ischia is an undetermined volcanic unit, probably corresponding to a volcanic acoustic substratum, eroded at its top by an unconformity, probably subaerial. The lower seismic unit of the basin filling is characterized by aggradational seismic reflectors. The intermediate first seismic unit of the basin filling is characterized by parallel to sub-parallel seismic reflectors. The intermediate second seismic unit of the basin filling is characterized by parallel to sub-parallel seismic reflectors. Forced regression and lowstand prograding wedges have also been distinguished. The hummocky deposits, characterized by a wedge-shaped external geometry and by a chaotic facies, have individuated during a main volcano-tectonic event occurring on the western Ischia offshore off Forio d'Ischia.

5.5.4. Seismic profile L31_6

The seismic profile L31_6 has been recorded from Punta Cornacchia (northern Ischia) to the canyon's system named the Punta Cornacchia canyon's head. It has been interpreted to show the active and recent erosional processes on the north-western Ischia continental shelf, focussing on the canyon system.

Several seismic units have been recognized based on the geologic interpretation of the seismic profile (Figure 33). The volcanic unit of Punta Cornacchia is represented by an undetermined volcanic unit, probably corresponding to a volcanic acoustic substratum, eroded at its top by an unconformity, probably subaerial. The second seismic unit is represented by a relict Pleistocene prograding wedge. The lower seismic unit of the basin filling is characterized by parallel to dipping seismic reflectors. A canyon basin filling has also been distinguished. A marine seismic unit cropping out at the sea bottom is characterized by interstratified fossil channel-levee complexes.

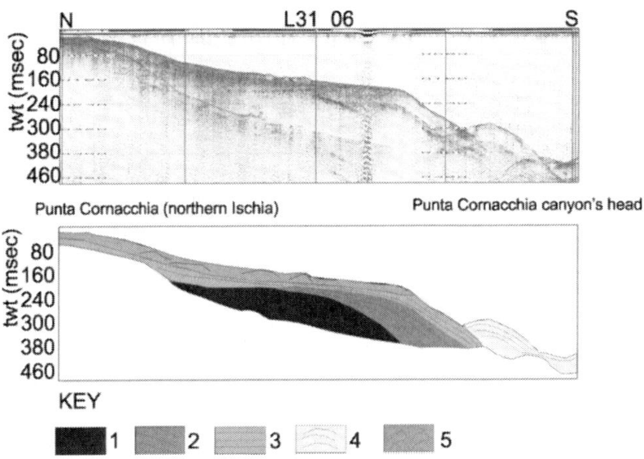

Figure 33. Seismic profile L31_6 recorded from Punta Cornacchia (northern Ischia) to the Punta Cornacchia canyon head (northern Ischia offshore) and corresponding geologic interpretation. Key. 1. Volcanic unit of Punta Cornacchia. Undetermined volcanic unit, probably corresponding to a volcanic acoustic substratum, eroded at its top by an unconformity, probably subaerial. 2: Pleistocene relict prograding wedge. 3: Lower seismic unit of the basin filling, characterized by parallel to dipping seismic reflectors. 4: Canyon basin filling. 5: Marine seismic unit cropping out at the sea bottom, characterized by interstratified fossil channel-levee complexes.

The stratigraphy offshore north-western Ischia (Punta Cornacchia) is characterized by an undetermined volcanic seismic unit (volcanic unit of Punta Cornacchia), probably related to the Zaro lava flow. The volcanic unit is overlain by a Pleistocene relic prograding wedge. The lower seismic unit of the basin filling is characterized by parallel to dipping seismic reflectors. A canyon basinal unit has also been identified. Finally, a marine seismic unit cropping out at the sea bottom is characterized by interstratified fossil channel-levee complexes.

5.5.5. Seismic profile L32_6

The seismic profile L32_06 has been recorded from Punta del Soccorso (western Ischia) to the north-eastern Ischia offshore (Figure 34). The scientific themes carried out through the interpretation of seismic profiles regard the debris avalanche deposits offshore Punta del Soccorso (western Ischia; Budillon et al., 2003a; 2003b). A sketch stratigraphic diagram offshore western Ischia in correspondence to Punta del Soccorso has been constructed based on the geological interpretation of seismic profile.

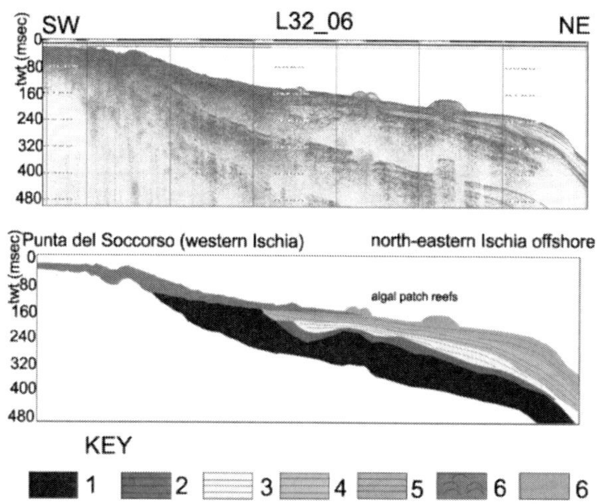

Figure 34. Seismic profile L32_6 recorded from Punta del Soccorso (western Ischia) to the north-eastern Ischia offshore and corresponding geologic interpretation. Key. 1. Volcanic unit of Punta del Soccorso. Undetermined volcanic unit, probably corresponding to a volcanic acoustic substratum, eroded at its top by an unconformity, probably subaerial. 2. Syntectonic sequence, locally filling the depressions at the top of the volcanic acoustic basement. 3. Lower seismic unit of the basin filling, characterized by progradational reflectors. 4. Intermediate seismic unit of the basin filling, characterized by progradational reflectors. 5: Upper seismic unit of the basin filling, characterized by progradational reflectors. 6: Hummocky deposits characterized by chaotic seismic sequences, offshore western Ischia (Punta del Soccorso). 6: Algal patch reefs.

Several seismic units have been distinguished based on the geologic interpretation of the seismic profile L32_6. The volcanic unit of Punta del Soccorso is an undetermined volcanic unit, probably corresponding to a volcanic acoustic substratum, eroded at its top by an unconformity, probably subaerial. A syntectonic sequence, locally filling the depressions at the top of the volcanic acoustic basement, has also been recognized. The lower seismic unit of the basin filling is characterized by progradational reflectors. The intermediate seismic unit of the basin filling is characterized by progradational reflectors. The upper seismic unit of the basin filling is characterized by progradational reflectors. Hummocky deposits characterized by chaotic seismic sequences, offshore western Ischia (Punta del Soccorso), overlain by algal patch reefs have also been recognized.

5.7. Sorrento Peninsula – Capri Island

From a physiographic point of view the Capri island represents the western elongment of the Sorrento Peninsula; both the sectors are characterized by the occurrence of Mesozoic-Tertiary successions having a carbonate platform facies. In particular, the Mesozoic-Tertiary carbonate succession having a platform facies, cropping out in the Sorrento Peninsula is replaced, in the Capri island, by a coeval succession of platform margin to proximal basin, characterized by the occurrence of stratigraphic gaps having a variable width.

The coasts of the Capri island have a total development of 26 km and are composed of steep coastal cliffs incised in calcareous-dolomitic and terrigenous units. Sometimes small actual beaches, composed of pebbles and coarse-grained sands occur at the toe of coastal cliffs or in small bays (pocket beach), where join the incisions eroding the carbonate reliefs, sometimes continuing in a subaqueous environment (D'Argenio et al., 2004; 2012).

New geological and geophysical data on the Capri island have been collected in the frame of the redaction of the new geological map of the Capri island at the scale 1:10.000 (D'Argenio et al., 2012). The marine cartography is based on the interpretation of geophysical data (bathymetric and seismo-acoustic) coupled to the analysis of sea bottom samples.

The Capri island is mostly composed of a Mesozoic-Tertiary carbonate succession, having a carbonate platform margin-basin facies, related to the western margin of a domain of carbonate platform. It is attributed to the western margin of a domain of carbonate platform (D'Argenio, 1976), to the

Campania-Lucania-Calabria carbonate platform (Sgrosso, 1986) or to the Apenninic platform (Mostardini and Merlini, 1986). Its eastward prolongation, individuated in the Lattari Mt. is composed of the S. Costanzo Mount-Punta Campanella. This sector of the Sorrento Peninsula is characterized by a thick calcareous-dolomitic succession having a carbonate platform facies and by the occurrence of a tectonic plan north-eastward verging, which determinates the doubling of the succession. The same structural pattern, also if referred to successions with a slope-basin facies can be observed along the northern coasts of the Capri island, where the tectonic superimposition of the carbonate succession on the Miocene (Langhian) siliciclastic deposits may be observed.

In particular, at Punta Sbruffo (Capri) it is possible to observe the superimposition of two carbonate thrusts: the upper one is composed of thick Ellipsactinia limestones, Middle Jurassic in age and the lower one by Oligocene cherty limestones. The present-day tectonic setting has been determined during the Pleistocene by the vertical extensional tectonics, causing the down throwing of the thrust surfaces. The counter-Apenninic (NE-SW) normal faults have mostly influenced the morphological setting of the island. The Apenninic (NW-SE) normal faults had a minor role, nonetheless the significant occurrence of the discontinuities having this direction.

The eastern sector (Capri) is morphologically lower with respect to the western sector (Anacapri), whose apex is represented by the Solaro Mt. The eastern sector has an articulated trending made of blocks; among the structural highs of the S. Michele Mt., Tuoro Mt. and Castiglione Mt. a relative low is individuated (D'Argenio et al., 2004; 2012).

The thinning of the island in its intermediate portion and the occurrence of the saddle which divides Capri from Anacapri should be explained by morphoselective processes provoked by the differential erosion among the siliciclastic successions, cropping out in correspondence to the Capri saddle and the carbonate ones, respectively occurring in the eastern and western sectors of the island (D'Argenio et al., 2004; 2012).

The coastal morphology, along the emerged sector, is characterized by articulated and steep areas alternating to beaches, located at the toe of fossil or inactive coastal cliffs and to active coastal cliffs located on calcareous-dolomitic promontories with small bays and pocket beaches. The trending of the shoreline and of the coastal cliffs is strongly conditioned by the tectonic lineaments (De Pippo et al., 2007; Ferranti and Antonioli, 2007), which, together with the denudational processes, have provoked the formation of the so called "Faraglioni di Capri" along the southern sector of the island

(Faraglione Grande, Faraglione di Mezzo, Scoglio del Monacone, Faraglione di Matermania).

The clastic littorals, having a grain-size mainly composed of gravels and sands are subjected to periodic processes of erosion and characterize a great part of the coastal sectors of Marina Grande and Marina Piccola in the Capri island; clastic littorals, usually having a reduced extension and mainly composed of gravels, occur in the restricted coastal bays, where there are many pocket beaches (Spiaggia di Punta del Capo, Torre Saracena, Marina Piccola, the Latino at Capri).

The remnant part of the coastal sector is represented by active coastal cliffs, having a subvertical development, mainly superimposed on carbonate lithotypes, often terrigenous (Capri). In some cases along the active coastal cliff detrital deposits occur, heterometric and heterogeneous, connected to palaeolandslides (Guadagno and Mele, 1994; Pellegrino, 1994), extending also in a submarine environment, often in the inner part or at the base of the numerous incisions eroded in an emerged sector, often prolonging up to 15 m of water depth (Donadio and Sgrosso, 2001).

The submerged coastal belt presents a particular interest due to the geoarcheologic, geomorohological and morphoevolutive characteristics and often due to the strong antropization of the shoreline. The localized archeological emerged and submerged hints of the Roman epoch in the Capri island (Grotta Azzurra, Bagni di Tiberio) have contributed, together with the morpho-stratigraphic data, to the reconstruction of the main phases of the recent evolutionary dynamics of these coastal belts (D'Argenio et al., 2004; 2012).

In the coastal belt included among 0 and 30 m, at the bottom of the bays (Cala Ventroso, Unghia Marina, Marina Piccola, Marina Grande) and at tracts along the coastal cliffs, at different water depths, marine abrasional surfaces, Pleistocene in age, have been submerged due to the Quaternary tectono-eustatic phenomena; sometimes these terraces are involved by deep incisions produced by the water courses during the periods of emersion representing the submarine prolongation of these occurring onshore, generating strict bays (Cala del Rio, Cala di Mezzo, Cala del Limmo). Moreover, along the submerged part of the vertical coastal cliffs occur at several water depths different cavities having a mixed nature, tectono-karstic and morpho-selective (Grotta Azzurra, Grotta Bianca, Grotta Verde); these latter ones, similarly to those ones of the Sorrento Peninsula (Romano and Sgrosso, 1992; Alessio et al., 1996; Donadio et al., 1996) are located at water depths ranging between 3 and 25 m.

The most important morphologic evidence which characterizes the sea bottoms is constituted by the continental shelf, which results well developed among the north-eastern sector of the Capri island and Punta Campanella in the Sorrento Peninsula and more reduced along the other sectors. Where the continental shelf is well developed in its main morphologic characters, it shows a low gradient and its rim (shelf break) is located at water depths varying among 80 and 150 m.

The sea bottoms of the western and south-eastern sectors reach even at a short distance from the coast water depths exceeding 500 m and are characterized by the prolongation in a submarine environment of the vertical coastal cliffs, whose thick foot often occurs at 60 m of water depths; starting from this water depth a gradual decrease of the slope is recorded. In the north-eastern sector (Bocche di Capri), instead, the sea bottom is flat up to Punta Campanella, not exceeding the 80 m of water depths.

The high part of the continental slope and often its rim seem to be incised by canyon's heads or morphologies, often fossil, linked to linear erosional processes in a continental environment or to denudational phenomena of a areal type connected to mass movements (collapses, slides, etc.). In the cases in which the continental shelf is more developed, characteristic morphologies induced by chaotic deposits linked to large-scale gravitational events are well developed (Marina Grande-Bocche di Capri).

The sectors of sea bottom in which the continental shelf results more limited and incised and then with a rim next to the tract of submerged coastal cliff have individuated in correspondence to Punta dell'Arcera – Punta Carena.

The structures of Capri and Sorrento Peninsula represent the emergent crests of a down thrown block, inclined of 7-10°, whose structural trends have configurated, starting from the Early Pleistocene, a half-graben structure, partly occupied by the Naples Bay and partly filled by marine and volcaniclastic sediments (Finetti and Morelli, 1974; Fusi, 1996; Fusi et al., 1991). The next geophysical surveys in the marine sectors have documented that the first phases of collapse and flooding of the Naples Bay have originated a wide and thick unit having a parallel stratification attributed to the Early Pleistocene (Aiello et al., 2001). This is in agreement with the chronological data furnished by the drill site Trecase 1 (Bernasconi et al., 1981; Aiello et al., 2010), testifying moderate continental clastic supplies and the tectonic down throwing of the southern sectors of the carbonate ridge, now down thrown to form the western offshore, bounding the Salerno Bay (D'Argenio et al., 1973; Ippolito et al., 1973; Bartole et al., 1984; Milia and Torrente, 1999; Sacchi et al., 1994). Regional geological studies have documented the occurrence of a regional fault (the Capri fault) having a counter-Apenninic trending, bounding to SE the Capri continental shelf, at a distance from the coast ranging among 0.13 and 1 km about (Finetti and Morelli, 1974; Latmiral et al., 1971). Due to the asymmetric structural configuration, the continental shelf, extending northwards of the Sorrento ridge reach a maximum width of about 20 kilometres, while southwards it is bounded by a few kilometers. In correspondence to the Bocca Piccola saddle it is controlled by the Meso-Cenozoic basement, widely cropping out.

The movements along the Pleistocene structural lineaments have controlled the formation of sedimentary basins, the distribution of sediments and the rate of clastic supply, invoking for a progressive deepening of the sedimentary environments, in which, in areas presently of basin, debris flow superimpose to shallow water successions. In the northern sector the stratigraphic structure of the margin consists of sedimentary wedges overlying the Meso-Cenozoic basement, which tend to become thinner in proximity to the Capri coasts (Latmiral et al., 1971).

Recent studies having a morphobathymetric and seismo-stratigraphic character on the sea bottoms between the Capri island and the Posillipo Cape have been recently carried out (de Alteriis and Toscano, 2003; D'Argenio et al., 2004; Aiello et al., 2011a). In detail for the emerged and submerged coastal belt several studies have been carried out starting from the '80, in which the geomorphologic, sedimentologic, geoarcheologic and morpho-evolutive aspects on the sea bottoms have been discussed (Cinque, 1986; Schiattarella, 1990; Barattolo et al., 1992; Cinque and Putignano, 1992; Romano and Sgrosso, 1992; Antonioli and Ferranti, 1996; Donadio et al., 1996; De Pippo et al., 2007).

Most part of the information on the morphology and the sedimentology of the coastal area of Capri refers to a preliminary work of geological mapping, in which the terrestrial information joined the marine information with the description of the main lithofacies associations occurring at the sea bottom (D'Argenio et al., 2004). In a sketch geological section and in a Sparker seismic line the geological characters of the Bocca Piccola saddle have been evidenced (D'Argenio et al., 2004). The authors describe wide terraced surfaces, developed during the last transgressive-regressive cycle, leaving a trace also with the formation of depositional terraces and abrasional surfaces located at several heights, whose chronological attribution is difficult to interpret. This study has contributed to the definition of both the main textural elements in the Capri island and to the morphological lineaments.

The sedimentological characteristics of the continental shelf and slope offshore the Capri island have been recently discussed (D'Argenio et al., 2012). As a general rule, the lacking of sedimentary covers in the Capri island is due to a low siliciclastic supply from adjacent mainland, whose contribution to the depositional environment of the continental shelf is due to small rivers characterized by reduced drainage basins. This condition favoured a scarce sedimentation and/or condensation, which determinates a selective enrichment in coarse-grained sediments, often having an organogenic composition up to high water depths (D'Argenio et al., 2012). Moreover, the strong action of the bottom currents, controlled by the oceanographic circulation and by the particular morphology of the Capri saddle, testified by the wide diffusion of current lineaments on the continental shelf, has determined the re-distribution to greater water depths of the fine-grained silty fraction. The Holocene covers thin proceeding seawards, having a constant thickness in the distal sectors, with a sedimentation rate typical of the hemipelagites.

On the continental shelf the facies associations with a scarce grain-size sorting prevail and the sands contain variable quantities of pelites. Nonetheless this, the more diffused sediments are represented by the organogenic and bioclastic sands, known in literature as the coastal detritic deposits, characterized by a biological content prevailing with respect to the clastic and lithic elements, often composed of fragments of organisms preserving the original pigmentation and indicating a scarce rate of rielaboration (D'Argenio et al., 2012).

These deposits are generally restricted to the more superficial levels of the sea bottom and of the first subbottom, being characteristic of a particular bathymetric interval.These deposits are in equilibrium with the present-day environmental conditions, overlying pelitic and sandy-pelitic transgressive sediment covers (D'Argenio et al., 2012). The organogenic deposits may be composed of variable rates of pelitic and sandy matrix. Moreover, for the occurrence of a belt of coarse-grained organogenic sediments, the natural decrease in grain-size and the relative increase in the pelitic fraction which may be observed with the increase of water depth, shows a significant inversion at water depths among 60 and 90 m. The diffusion of these lithofacies associations, including from the sandy gravels to the gravel pelites is particularly wide in the Capri island and involves the most part of the sea bottoms of Bocca Piccola and all the area surrounding the island up to water depths of 90-100 m (D'Argenio et al., 2012).

At greater water depths the pelitic sediments prevail, with a high rate of sorting, particularly in the sectors in proximity of the retreating canyon's heads and along the slopes in continuity with the coastal cliffs. The outer shelf deposits are involved by parallel lineations indicating the occurrence of bottom currents (D'Argenio et al., 2012). Many coralligenous patch reefs have also been observed at water depths up to 170 m.

In the sectors located northwards of the Capri island beach rock deposits have been observed, having a high backscatter (D'Argenio et al., 2012). The depositional characters derived from the analysis of sea bottom samples and by the interpretation of the acoustic images suggest that the sea bottoms offshore the Capri island are actually starved and a significant part of these sediments is produced *in situ* due to the intense biologic activity. This is favoured by a limited supply of nutrients from the emerged mainland, mainly composed of carbonate terranes, creating conditions for a low rate of eutrophization.

The marine geomorphology of the Capri island has been recently discussed (D'Argenio et al., 2012). The sea bottoms, in relationships to the local morphological configuration in the frame of the emerged sector of the Capri island show a set of particularities such to distinguish physiographic environments with similar morphometric and morphostratigraphic characters. It is strong the control of the lithostratigraphic and structural setting on the evolution of the submerged landforms and on the morpho-evolutive processes. A strong control on the physiography, structuration and evolution of the investigated sea bottoms is exercised from the coastal typology, characterized by high coastal cliffs and by the absence of low coasts (D'Argenio et al., 2012).

The high coastal cliffs individuates in the sectors characterized by slopes directly adjoining the sea, with high gradients independent from the lithology of the cropping out units. The emerged coastal cliff prolongates in a submarine environment and develops with similar characteristics, generally up to the isobaths of 5-10 meters, where they join the flat sea bottom. The last one develops with homogeneous characteristics, showing often some irregularities due to buried structures and morphologies, not often identified and characterized from a geologic and geomorphologic point of view. Sometimes the submerged coastal cliffs prolongate to 30-50 m of water depth (D'Argenio et al., 2012).

The submerged coastal cliffs develop according to two different morphological types: the first one is represented by sub-vertical to vertical walls; the second one is characterized by a relief which differs from the high-angle trending, up to degrade with average values of slope, usually with an articulated morphology due to the occurrence of terraced surfaces, detritic bodies coming from the dismantling of the emerged coastal cliff, channels and morphological slopes.

In some cases, along the emerged coastal sectors, to tracts of high coasts not directly adjoining the sea for the occurrence of narrow belts of beaches corresponds a tract of submerged beach and of a narrow shelf degrading towards the slope rim (depositional terraces) located at some hundred of meters of distance proceeding seawards (D'Argenio et al., 2012).

The geomorphology of the Capri island is characterized by the occurrence of rocky outcrops mainly nearshore, having a carbonate nature and different ages. These outcrops are formed by the seaward prolongation of the coastal cliffs or by blocks having also great dimensions. The highest coastal cliffs reach water depths highest than 50 m, mainly in the north-western sector of the Capri island (D'Argenio et al., 2012).

The remaining tracts of sea bottom are characterized by wide sedimentary covers, mainly composed of sandy deposits with a middle to fine-grained texture, sandy gravels and pebbles having a carbonatic nature (D'Argenio et al., 2012). The pelitic deposits are less diffused, together with the blocks and the rocky outcrops isolated seawards. The occurrence of marine Phanerogams is evidenced by wide Posidonia grasses at water depths ranging from 5 to 25 m (D'Argenio et al., 2012).

The morphostratigraphic and structural characteristics of the sea bottoms in the coastal marine area of the Capri island are linked to its tectonic and eustatic evolution and to the control of this latter one on several morphogenetic processes which are developed along the shoreline. A first morphostructural characterization of the investigated sea bottoms is evidenced by the analysis of the morphobathymetric trending, showing a strict correlation with the emerged sector. In correspondence to the high coastal cliffs a great variety of landforms occur, due both to the tectonics and to the alternance of erosional and depositional processes. Other important control on marine geomorphology are represented by the continuous erosional processes and the landslide events (D'Argenio et al., 2012).

Geologic and geomorphologic evidence shows a complex evolution of the coastal cliffs. The main morphostructural lineaments shown by the bathymetric trending individuate fault systems having a NE-SW, NW-SE and E-W trending, according to those observed onshore (Acocella and Funiciello, 1999; Acocella et al., 2004). In the Capri island the thrust between thick Mesozoic limestones and Miocene flysch has also been observed. All these structural elements have controlled the evolution of the coastal cliff, with the development of bays superimposed on normal faults and by the rectilinear trending of the shoreline (D'Argenio et al., 2012).

The high grade of fracturing of the carbonate lithotypes, together with the erosion at the toe of coastal cliff due to wavy motion, are partly responsible for the instability processes actually observed. The general trending of the recognized morphostructural elements are in agreement with the fracture systems observed in carbonate rocks (Schiattarella, 1990).

In the whole coastal sector of Capri island relict morphologies connected to old stands of the sea level lower than the actual one or in a way of formation. The overall framework of the geomorphological data evidences the occurrence of a set of marine morphologies such as abrasional terraces and corrosion notches, more or less pronounced starting from the average sea level up to – 15 m (D'Argenio et al., 2012).

The terraced surfaces, having a significant extension and well recognized from a morphological point of view have been observed at water depths of 3 m, 5/6 m, 8/10 m, 12/15 m, 20/21 m, 26 m and 32 m about with respect to the present sea level (D'Argenio et al., 2012). These surfaces are superimposed mainly on conservative lithologies, mainly on the carbonate lithotypes. The occurrence of main gullies indicates a subaerial modeling during phases of emersion due to tectono-eustatic causes (D'Argenio et al., 2012).

Some wide depositional terraces have been individuated through the morphobathymetric analysis, starting from 15/20 m in the area joining with the inner platform. At these water depths coarse-grained deposits have been observed, located at the toe of structural slopes and morphological highs and perhaps corresponding to palaeoshorelines related to sea level lowstands lower than the present one (D'Argenio et al., 2012). Seven orders of terraced surfaces have been observed in the Capri island.

The geomorphology of the continental shelf and slope have also been discussed (D'Argenio et al., 2012). The topography of the Capri offshore is strongly controlled by the occurrence of the acoustic basement, cropping out at the sea bottom and by the Pleistocene glacio-eustatic variations, which have determined the occurrence of transgressive-regressive cycles with a high morphoevolutive potential.

The most important morphological element is represented by the continental shelf, whose trending is strongly controlled by the structural asymmetry of the Mesozoic carbonate block Capri-Sorrento Peninsula and by the Middle and Late Pleistocene sedimentary covers, accumulated in the down thrown sector located northwards of the structure. The southern sector of the continental shelf is more reduced, since it is individuated in correspondence to the Capri master fault (Bartole, 1984; Sacchi et al., 1994; Aiello et al. 2009; 2011a; D'Argenio et al., 2004; 2012). Wide portions of the continental shelf seem to retreat; here the shelf break has the typical shape with a seawards concavity and is located at variable water depths as a function of the entity of retreatment, up to – 80 m. In these sectors the continental slope is incised by several channels (Aiello et al., 2009a). This morphological aspect is typical of the slope evolving through channellised fluxes, characterized by the occurrence of gullies, which at water depths merge together forming submarine canyons. In correspondence to the high coastal cliffs of Punta Carena and among the Faraglioni di Capri and the "Il Monacone, the continental shelf is practically lacking, if we exclude limited remnant rims with a shallow shelf break (Aiello et al., 2009c).

Limited outcrops of acoustic basement interrupt the regularity of the sea bottoms and have been observed, in particular, to the south-west of Punta Carena and along the margin of the continental shelf, up to the south of Punta Campanella. They represent some morphological relicts preserved from the erosional retreatment of the continental shelf (D'Argenio et al., 2012).

To the west and to the north of the Capri island the continental shelf is extended and widens with average gradients among 0.3 and 0.5 degrees. Its rim is located at 3/5 km from the shoreline. A concave break in slope has been instead observed at water depths rangig between 110 and 130 m and probably corresponds at the foot of littoral bodies formed during the last stage of relative sea level lowstand, overlying with a stratigraphic contact on palimpsest erosional surfaces of the previous glacio-eustatic cycle and on portions of outcropping acoustic substratum (D'Argenio et al., 2012).

Several discontinuities to the regularity of the submarine topography of the outer shelf are represented by small reliefs having a subcircular shape, located at water depths ranging among 100 and 150 m (D'Argenio et al., 2012). These structures have been interpreted as coralligenous bioconstructions; some of them should be still active. They are represented by small reliefs at the sea bottom and have been interpreted in the geological maps (D'Argenio et al., 2012) as organogenic lithic bodies, formed by benthic organisms (bryozoans, corals, echinoids) and surrounded by coarse-grained bioclastic sediments encrusted by calcareous algae, Mollusks and foraminifera, derived from the dismantling of the benthic organisms (D'Argenio et al., 2012).

The platform to the north of Capri island is characterized by wide outcrops of chaotic bodies formed by blocks reaching dimensions of several meters and large slide bodies, determining an abrupt topography between 50 and 120 m of water depths. These bodies, fallen by retreating and erosional coastal cliffs, are presently immersed in a coarse-grained detritic matrix and are partially draped by actual and sub-actual hemipelagic deposits at high water depths and by organogenic deposits at water depths included among 60 and 110 m. They are physically separated from the present coastal cliffs due to the progressive retreatment from the shoreline. Active and recent slide bodies having a reduced extension overlie the present-day coastal cliffs (D'Argenio et al., 2012).

The slope along the north-western and western sectors of the Capri island joins the continental shelf to the basinal areas of the Naples Bay crossed by the Dohrn canyon (Aiello et al., 2001; D'Argenio et al., 2004; Milia, 2000; Di Fiore et al., 2011).

The sea bottoms of the Capri island have been recently investigated and their sediments mapped through marine geological mapping at the scale 1: 10.000 (Geological Map n. 484 Isola di Capri, scale 1:10.000). The main deposits reported in the geological map (marine area) are here briefly resumed. The Quaternary deposits are represented by coastal defence works, landslides, beach deposits and eluvial-colluvial blankets. The coastal defence works are represented by carbonate or lavic heterometric blocks up to few meters in size used for emerged and submerged coastal defense works. The landslides are gravitational deposits still under accumulation, with heterogeneous lithoclasts, chaotically mantling the toe of the slopes. The chaotic deposits from slow to fast sliding of volcaniclastic deposits mantle the carbonate slopes and collapse landslides along the coastal carbonate scarps at the base of the steeper slopes. The beach deposits are composed of sandy deposits, medium and coarse sands with large carbonate blocks due to undermining of the coastal cliffs owing to marine erosion, at present located in emerged beach sectors. The eluvial-colluvial blankets are represented by reworked pyroclastic deposits, mixed with carbonate debris, often interbedded with more or less thick pyroclastic levels or thin pumice beds, variously degraded. These deposits range in age from the Late Pleistocene to the Holocene.

The deposits of littoral environment include the toe of coastal cliff deposits and the submerged beach deposits. The toe of coastal cliff deposits are composed of heterometric lithoclasts, mainly carbonatic, from gravel to block in size, generally sharp-edged, locally rounded. They are subdivided into two lithofacies units on the base of textural parameters. The first unit is represented by the blocks. Heterometric blocks vary in size from decimetries to decametries, with sharp or poorly smoothed corners, made of limestones and dolomitic limestones in the local succession in scarce or absent gravel matrix. The second unit is represented by the gravels and pebbles, such as heterometric gravels of heterogeneous texture, composed of limestone clasts, from rounded to sharp and varying in size from centimetric to decametric. The submerged beach deposits are lithoclastic deposits mainly carbonatic in origin, frequently with bioclasts, mostly Mollusks. The clasts are generally well sorted, from rounded to sub-rounded. The matrix is scarce or absent. The submerged beach deposits are formed by different lithological units according to the hydrodynamic regime and wave regime. The lithological units are the litho-bioclastic gravels and sands, the litho-bioclastic sands and gravels and the litho-bioclastic sands. The litho-bioclastic gravels and sands are composed of gravels, sandy gravels and coarse-grained sands with bioclasts, pebbles and rare limestone blocks, from rounded to sub-rounded, in a scarce middle-fine grained sandy matrix. The litho-bioclastic sands and gravels are composed of coarse to middle-grained sands, sometimes with gravels; with bioclasts and including carbonatic pebbles. The pelitic matrix is scarce or lacking. The litho-bioclastic sands are composed of middle to fine-grained sands, sometimes with carbonatic fragments. The pelitic matrix is scarce or lacking. These deposits range in age from the Late Holocene to the present.

The deposits of continental shelf environment include the shelf deposits and the bioclastic deposits. The continental shelf deposits are composed of both lithoclasts and bioclasts and are subdivided into different lithological units on the base of hydrodynamic regimes and sedimentary components. The lithological units are the blocks, the blocks and pebbles, the sands and gravels, the litho-bioclastic sands, the pelitic litho-bioclastic sands and the sandy mudstones and mudstones. The block unit is composed of heterometric blocks varying in size from decimetric to decametric, with sharp or poorly smoothed corners, made of limestones and dolomitic limestones of the local succession in scarce gravelly matrix sometimes absent. The blocks and pebbles unit is composed of heterometric blocks and pebbles, varying in size from decimetric to decametric. The unit is carbonatic in origin and shows a scarce gravelly matrix. The sands and gravels unit is composed of medium to coarse-grained sands, mainly bioclastic, in a scarce pelitic matrix. The litho-bioclastic sands unit is composed of sands, fine to medium, litho-bioclastic. The main components are made of carbonatic fragments, with variable percentage of bioclasts in a scarce pelitic matrix, sometimes organogenic. The pelitic components is more abundant in the distal areas. The pelitic litho-bioclastic sands are formed by carbonatic and bioclastic sands, sometimes pyroclastic in nature. The pelitic matrix is scarce or lacking. The bioclastic deposits are formed by different lithological units related to the composition and texture. They are represented by the organogenic sands and gravels, by the biodetritic coarse-grained sands, with scarce pelitic matrix, containing benthic communities of the coastal detritic assemblage; mud supported organogenic gravel and bioclastic sand with benthic communities, typical of the muddy coastal detritic assemblage; subordinately pelitic sands, sands and fine gravels, originated in situ from the breakdown of coralgal mounds. The age of these deposits range from the Holocene to the present.

The slope environment deposits are characterized by the continental slope deposits, represented by pelites and pelites with a variable fraction of bioclastic and volcaniclastic fine sands, fairly abundant along the canyon heads and locally, by pelites with sands and gravels. Their age is Holocene.

Several pre-existing units have also been distinguished in the marine area (Geological Map n. 484 Isola di Capri, scale 1: 10.000). The Late Pleistocene relict deposits are formed by pelitic sands and sandy mudstones with bioclastic and volcaniclastic fragments. Another type of Late Pleistocene relict deposits are formed by breccias and blocks, sometimes in a scarce gravel matrix. The Marina Piccola breccias are unsorted edgewise breccias with moderately cemented to loose calcareous elements and variable amount of pyroclastic matrix, normally unflected by tectonics. These breccias form loose talus, variable in thickness (only locally exceeding 20 m), at the foot of the slopes. The Tragara unit is composed of marine calcareous puddingstones in a loosely cemented calcarenitic matrix. The elements are well rounded and varying in size from a few centimetres to several decimeters. The Tragara unit is thick less than one meter. The Damecuta unit is composed of marine calcareous puddingstones and calcarenites moderately to well cemented, with rounded elements of variable size in arenitic matrix and an overall thickness not exceeding a meter.

A detail of the multichannel seismic profile Sister 4_2 (Aiello et al., 2011a) has been constructed in order to show the seismic stratigraphy of the Bocca Piccola saddle (Capri island) towards the Naples Bay (Magnaghi canyon). The seismic profile is shown in Figure 35. The Bocca Piccola saddle is characterized by the occurrence of a seismic unit with a scarce penetration of the seismic signal (MC unit), correlated to Meso-Cenozoic carbonates and representing the acoustic basement of the Naples Bay. The MC unit is overlain by three seismic units (A, B and C units; Figure 35). The A unit is a NW-dipping progradational unit with eroded topsets and preserved clinoforms. The unit shows alternating intervals of highly continuous obliquous to parallel reflectors and acoustically transparent intervals. It has been interpreted as an Early-Middle Pleistocene relict prograding wedge. The B unit is a wedge shaped transgressive unit, developed in outer shelf and slope settings. It is characterized by highly continuous sigmoidal to obliquous clinoforms, interpreted as Pleistocene deposits occurring in the central part of the basin. The C unit is a progradational unit with well preserved offlap breaks thickening seawards. The unit laterally grades to the volcanic units Mg and Isch and is deeply incised in correspondence to the Dohrn canyon. It is represented by a Late Pleistocene prograding wedge supplied by the Sarno river mouth and by the Sorrento Peninsula tectonic uplift.

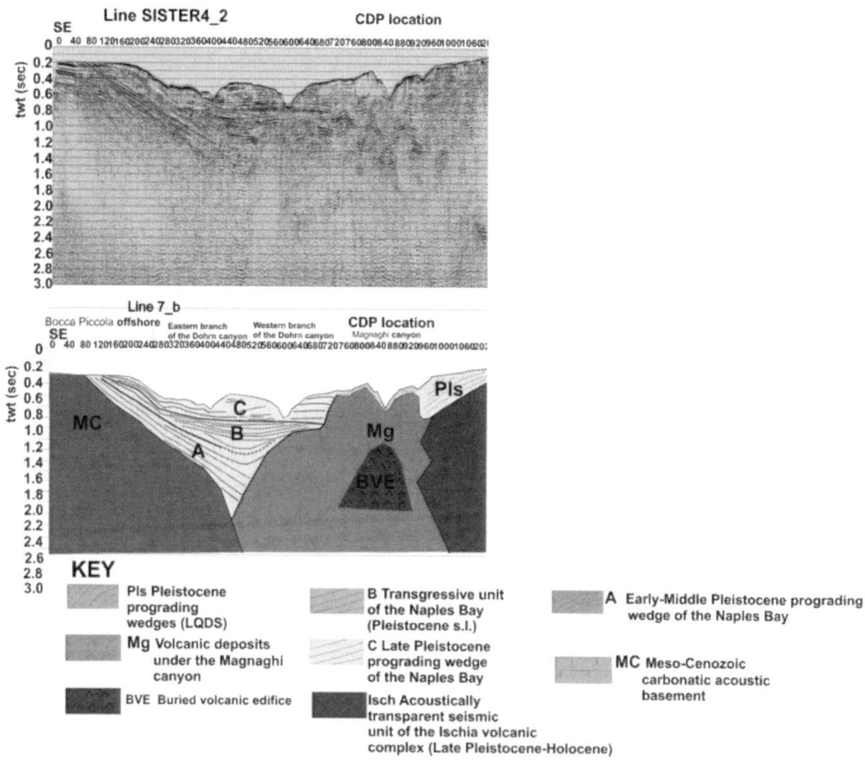

Figure 35. Detail of the seismic profile Sister4_2 showing the seismic stratigraphy of the Bocca Piccola saddle (Capri island) towards the Naples Bay (modified after Aiello et al., 2011).

A detail of the seismic profile Sister 9_1 (Aiello et al., 2011a) has been constructed in order to show the seismic stratigraphy of the Capri structural high towards the Salerno Valley (Gulf of Salerno). The Capri-Sorrento master fault, down throwing the Meso-Cenozoic carbonates of the Sorrento Peninsula-Capri island structural high seawards has been crossed by the seismic profile, shown in Figure 36. The Capri structural high is bounded by the Banco di Fuori structure to the north-west and by the Salerno Valley to the south-east through regional normal faults. The regional normal faults are respectively represented by the Dohrn fault and by the Capri-Sorrento fault, stronlgly down throwing the Meso-Cenozoic carbonatic unit, representing the deep bulk of the Naples Bay. The Capri structural high is characterized by the occurrence of the Meso-Cenozoic carbonatic unit (MC unit), characterized by an acoustically transparent facies with low penetration of the seismic signal. The unit is overlain by two thick seismic units, respectively the A and C units. The A unit, characterized by obliquous prograding clinoforms, represents an Early Pleistocene prograding wedge and is a lower unit in the stratigraphic architecture of the Naples Bay. The C unit is a Late Pleistocene prograding wedge representing an upper unit in the stratigraphic architecture of the Naples Bay. It is characterized by low angle sigmoidal to obliquous clinoforms, supplied by the palaeo-Sarno river mouth.

The morpho-bathymetry of sedimentary basins based on Multibeam surveys represents a research line of increasing interest, mainly for its implications in the coastal and deep sea environmental monitoring in terms of the definition of geologic and environmental hazard. The Multibeam bathymetric data collected during the oceanographic cruise Sister2 (Aiello et al., 2009c) have evidenced the high gradients of the continental slope and the occurrence of erosional processes, partly still active, mainly at the toe of the slope, where a dense network of channels occurs (Figure 37). The analysis of seismic profiles has already evidenced the tectonic activity of the fault escarpment as a triggering factor for gravity instabilities active during the Late Pleistocene and the Holocene; this is evidenced also by slumping deposits, characterized by a chaotic seismic facies, intercalated at several stratigraphic levels in the basin filling. On the contrary, in the distal zones of the valley, where deposition prevails, the shallower sector appears in a regime of intense erosion (Salerno canyon) and shows recent tectonic deformations, as antiforms and normal faults, which can be interpreted in the regional framework as hints of intense erosional tectonics, lasting up to recent times (Aiello et al., 1997a; 1997b).

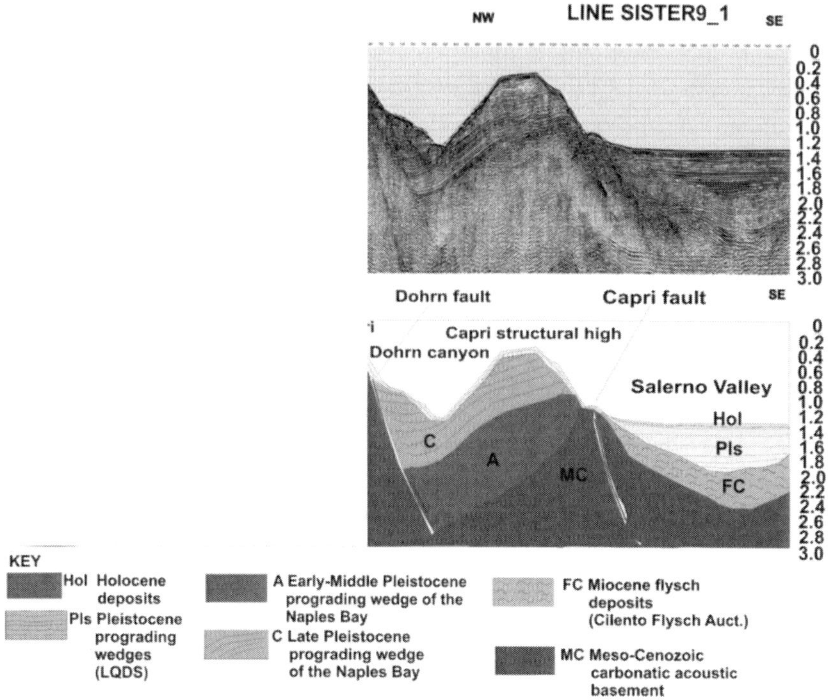

Figure 36. Detail of the seismic profile Sister9_1 showing the seismic stratigraphy of the Capri structural high towards the Salerno Valley in the Gulf of Salerno (modified after Aiello et al., 2011).

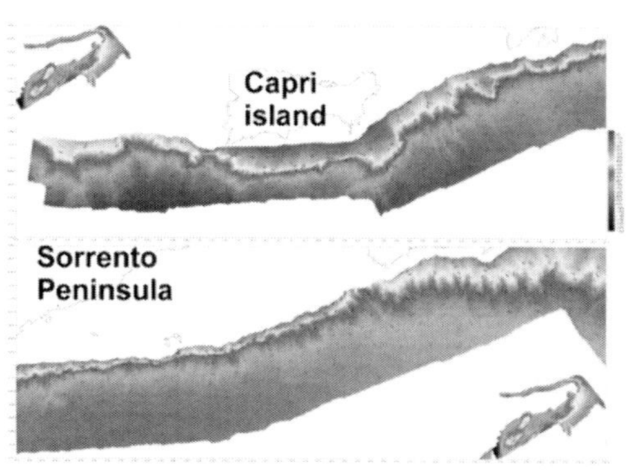

Figure 37. Contour map of the Multibeam bathymetry on the steep, tectonically controlled continental slope, located between the Capri island and the Sorrento Peninsula. It is irregularly articulated and is characterized by promontories and embayments, reflecting the main structural trends occurring in the Capri island.

The Multibeam bathymetric data have enabled the construction of a three-dimensional DTM reconstruction, covering an area of 1600 square kilometres, comprising the area between the southern coast of the Capri island to the north-west and the mouth of the Sele river to the south-east (Aiello et al., 2009). The continental shelf of the Salerno Gulf shows a strong asymmetry proceeding from north to south, reflecting the different structural domains of this sector of the Apenninic chain. The variability of extension, of the depth of the shelf break and of the average slopes are controlled by the structural and geological setting of the margin areas (Bartole et al., 1984), more than by the glacio-eustatic variations during the Pleistocene. The continental slope surrounding the Salerno Gulf is characterized by structural depocentres with a NW-SE Apenninic trending, originated by the extensional phases related to the Tyrrhenian basin, alternating with morphostructural highs, adjacent to intra-slope basins. In particular, the geological interpretation of the DTM has evidenced the occurrence of two complex NE-SW trending morpho-structural highs (Aiello et al., 2009c).

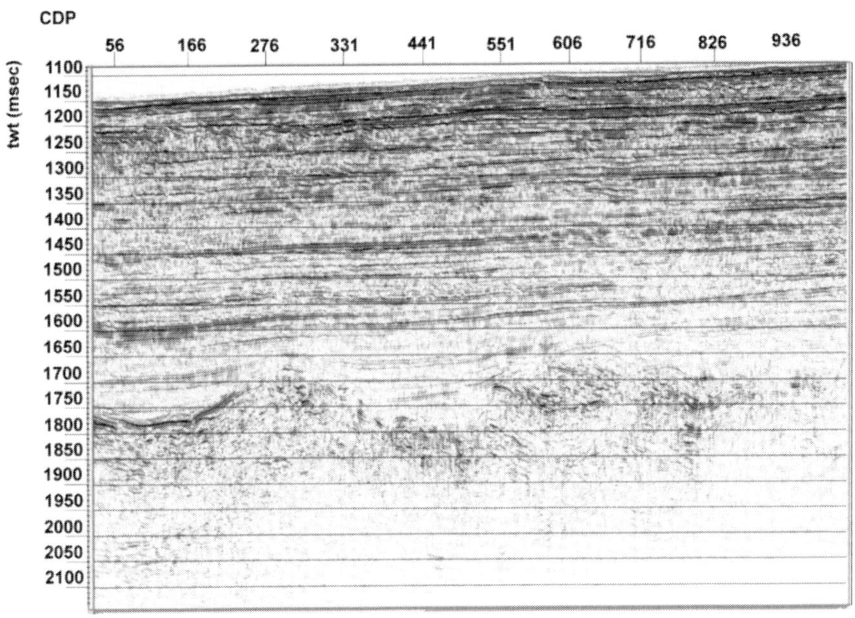

Figure 38. Detail of the multichannel seismic profile L6 showing the Meso-Cenozoic unconformity, located at the top of Meso-Cenozoic carbonates.

The DTM interpretation has already evidenced that the continental slope is a main morphologic and tectonic structure, NE-SW oriented; it starts to the west from the border of the Dohrn canyon, skirts the Capri island in a E-W direction and finally it is parallel to the Sorrento Peninsula, for an overall length of 12 kilometres (Aiello et al., 2009c). The slope is irregularly articulated and is characterized by promontories and embayments, reflecting the main structural trends occurring on land in the Capri island. Offshore the Capri island the slope is located at water depths ranging between 120 m and 1000 m and its gradients range from 17.8° to 23°. At the south-western promontory of the island (Punta Ventroso), the gullies occurring on the continental slope create a sub-circular sedimentary structure with a diameter of about 1500 m. In correspondence to the western rim of the Sorrento Peninsula (Punta Campanella) the continental shelf is very narrow (about 400 m); the slope starts at 120 m of water depth and stops in correspondence to a large slope fan.

A detail of the multichannel seismic profile L6 showing the Meso-Cenozoic unconformity is shown in Figure 38. Particular attention has been paid to the geological interpretation of the multichannel seismic profile L6, on which an important unconformity has been identified (Figure 38). The erosional unconformity, located at depths ranging between 3000 and 3500 meters (Figure 38) is located at the top of the acoustic basement and is strongly downthrown in correspondence to normal faults. The unconformity is interpreted as the top of the Meso-Cenozoic carbonatic sequence, widely cropping out in the adjacent emerged sector, in correspondence to the Sorrento Peninsula structural high. A thick basin filling, pertaining to the Salerno Valley, develops on the unconformity, having an average thickness of about 3000 m, probably Pleistocene in age.

CONCLUSION

The sea bottoms of the Naples Bay have been studied through geophysical data, including Multibeam bathymetry and high resolution seismic profiles. Several morphological and depositional settings have been distinguished, including the Naples Bay, the Vesuvius offshore, the Ischia island and the Sorrento Peninsula/Capri island physiographic domain. Several conclusions are here analysed to improve the geological knowledge on the Naples Bay.

Naples Bay

One goal of this chapter was to adopt an approach to the study of marine geophysical datasets based on the integrated interpretation of high resolution seismic and magnetic data. This method, even if qualitative in the present chapter, has improved the geological knowledge in an active volcanic area as the Gulf of Naples, since it has disclosed and located several main volcanic bodies, detected through the seismic interpretation and constrained through significant magnetic anomalies.

The emplacement of volcanic bodies in the subsurface of the Gulf of Naples has been controlled by the eruptive phases of Somma-Vesuvius, Phlegrean Fields and Ischia and Procida islands. The geological setting of the Gulf of Naples has also been controlled by the Dohrn canyon, acting as a main morpho-structural lineament in the investigated area, as confirmed by our data analysis. The Dohrn canyon western branch sharply separates a southeastern sector of the gulf, characterized by the occurrence of sedimentary units by a northwestern one, where main volcanic and volcaniclastic units are localized.

Weak magnetic anomalies have been recognized in the eastern part of the Gulf of Naples, offshore of the Somma-Vesuvius volcanic complex, where a shallow continental shelf occurs. Seismic interpretation has allowed the identification of submerged or buried parasitic vents, located offshore the Torre del Greco town. These vents are genetically related to the eruptive activity of the Somma-Vesuvius volcano during recent times (Aiello et al., 2005). One of these vents extends for almost 7.5 km and corresponds to a main magnetic anomaly, in the order of about 500 nT. The age of this vent could be only hypothesized, but it could be historical, taking into account both the reduced thickness of the Late Pleistocene sedimentary cover and the onshore volcanology (Vezzoli, 1988). The volcanological map of Somma-Vesuvius (Vezzoli, 1988) has shown how on the flank of the volcano facing on the Tyrrhenian sea, lava flows before 1631 and after A.D. 79 eruptions, frequently covered by thick pyroclastic deposits (glowing clouds, ash clouds, lahars) of the A.D. 1631 eruption extensively crop out.

From the Sorrento Peninsula to Naples town a wide seismic unit correlates with the Campanian Ignimbrite pyroclastic flow deposits (37 ky B.P.) and is unrelated with significant magnetic anomalies. On the contrary, weak magnetic anomalies are related with dome-shaped magmatic structures located at the top of the above mentioned seismic unit. These relatively small magmatic bodies are interstratified into Campanian Ignimbrite pyroclastic deposits produced by extensive explosive activity of the Phlegrean Field volcanic complex (Aiello et al., 2005).

The volcanic bodies recognized in the offshore of Somma-Vesuvius are interpreted as a strip of eruptive vents, settled on a system of NNE-SSW trending normal faults. These vents are associated with a magnetic anomaly field composed of several anomalies and reaching a maximum intensity of 500 nT. The largest of these structures shows three main elevated peaks, corresponding to several magnetic maximums in the maps of total magnetic field of the Naples Bay (Aiello et al., 2005). The preferential alignment of magnetic anomalies and related seismic structures along a NNW-SSE trending suggest such a structural elongment offshore the volcano, never singled out by previous papers, which evidenced systems of NE-SW normal faults (Bernabini et al., 1973; Finetti and Morelli, 1973; Cassano and La Torre, 1987; Bruno et al., 1998; Simini et al., 2003; Judenherc and Zollo, 2004).

Ischia Island

A sketch qualitative chronostratigraphic diagram of the Ischia volcanic deposits has been constructed, taking into account the stratigraphic relationships with the volcanic deposits pf Procida island (Figure 39). The hiatus in the sedimentation have also been shown. The volcanic deposits of the first eruptive cycle (135-100 ky B.P.) are represented in violet on the left in the diagram. The volcanic deposits pertaining to the Vivara, Fiumicello, Pozzo Vecchio and Terra Murata volcanoes have been represented on the right in the diagram with a dark orange colour. The Green Tuffs of the Epomeo Mount (55 ky B.P.) have been represented on the left in the diagram with a dark green colour. The volcanic deposits of the 2nd Ischia eruptive cycle (55-33 ky) are represented on the left in the diagram with a green colour. The volcanic deposits of the 3rd Ischia eruptive cycle (33-28 ky B.P.) are represented on the left in the diagram with a red colour. Breccia Museo, Campanian Ignimbrite and Solchiaro volcano (Procida eruptions) are represented on the left in the diagram. The Ischia volcanic deposits of the 4st eruptive cycle are represented on the left in the diagram. The Ischia volcanic deposits of the 5st eruptive cycle are represented on the left in the diagram.

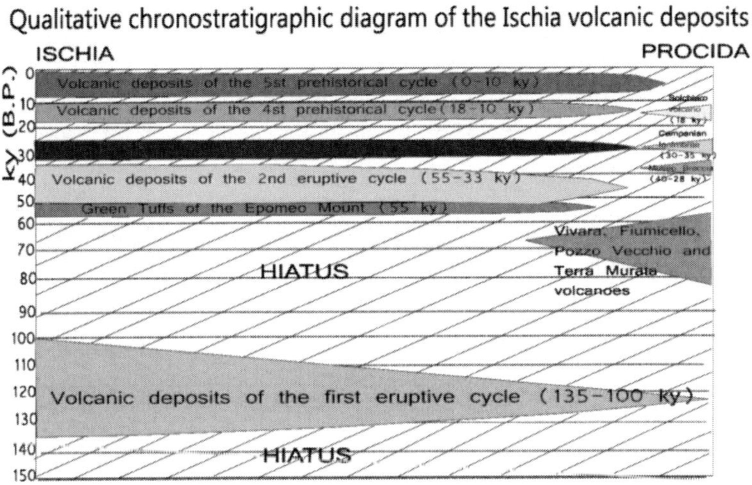

Figure 39. Sketch chronostratigraphic diagram of the Ischia island.

Four main phases in the eruptive activity of Ischia island are here resumed (Chiesa et al., 1985; Poli et al., 1987; Aiello et al., 2012a). The first eruptive phase is older than 150 ky B.P. and includes the emplacement of pyroclastic products and intercalated lava flows, without a clear origin, originated from single and multiple volcanic sources. The second eruptive phase range between 150 and 75 ky B.P. During this phase the extensional tectonics gave rise to the growth of several lava domes and to the production of pyroclastic materials. The semicircular distribution of the domes seems to indicate the occurrence of a caldera rim. These first two phases suggest an ancient volcanic complex, probably much larger than the present Ischia island and now almost completely dismantled, predating the eruption of the ignimbritic tuffs of Epomeo Mount. The third eruptive phase ranges between 55 to 20 ky B.P. and includes the emplacement of the Epomeo Green Tuff, an ignimbrite which covered the coastal relief and filled the depression at the center of the Ischia island. The vent erupting the Epomeo Green Tuff was located in the southern sectors of the island. During this phase the eruptive activity continued in correspondence of the volcanic centers originating the Citara-Serrara Fontana Formation (Rittmann, 1930). The fourth eruptive phase ranges between 10 ky B.P. and 1302 A.D. and the activity concentrates mainly in correspondence to the Ischia graben. The Ischia graben was controlled both by the faults generated during the rapid uplift of the volcano-tectonic horst of Monte Epomeo and by counter-Apenninic (NE-SW) normal faults. During this eruptive phase the Zaro lava flow (1302 AD) was the only eruption outside the

Ischia graben. This phase is characterized by lava flows and a minor fall-out of pyroclastic deposits from monogenic volcanoes.

The main seismic sequences of the north-western and south-eastern Ischia offshore have been reconstructed through the geological interpretation of seismic profiles. Regional seismic sections have been assembled based on their geological interpretation to improve the understanding of the structural and stratigraphic characteristics of the island offshore, taking into account the volcanology and the stratigraphy of the onshore sequences (Vezzoli, 1988; Orsi et al., 2004; Brown et al., 2008).

Offshore the Punta Imperatore - S. Angelo sector the Ischia continental shelf is very narrow and the Late Quaternary deposits are thin and restricted. The shelf break seems to be depositional. The most important seismic unit is the volcanic acoustic basement, cropping out on the continental slope under a thin drape of Holocene deposits and genetically related to the Punta Imperatore lavas.

At the Mavone cave offshore the volcanic acoustic basement, cropping out on the marine south-western slope of Ischia seems to be very thick and genetically related to the Mavone cave lavas. On the continental shelf the highstand deposits seem to be well developed and the shelf break is depositional. At the base of the Mavone cave promontory alkalitrachytic lavas dated at about 28 ky B.P. extensively crop out (Orsi et al., 2004). At the coastal cliff of the Mavone cave a stratigraphic section of the Scarrupo di Panza volcano crops out. It consists of a thick sequence of welded scorias, while at the Mavone cave (towards south) and at Punta Imperatore (towards north) the above mentioned lavas include levels of pumices and scorias.

Offshore Punta del Chiarito – S. Angelo the continental shelf is wider than in the adjacent sectors. Transgressive and highstand deposits occur. The transgressive deposits unconformably overlie an erosional truncation located at the top of a thick stratified unit corresponding to a prograding wedge. This unit overlies the volcanic acoustic basement, genetically related to the lava dome of Punta del Chiarito or to the S. Angelo eruptive center.

The Maronti littoral has been studied through three seismic sections perpendicular to the shoreline and one parallel. Here the continental shelf is very narrow with respect to the adjacent sectors. The highstand and transgressive deposits are thin and restricted. The volcanic acoustic basement is thick and continuous from the continental shelf towards the slope. Several retreating canyon's heads have been identified.

The Barano Bay has been studied through three seismic sections perpendicular to the shoreline. A thick volcanic acoustic basement has been recognized in correspondence to the slope named the Scarrupata di Barano. On the lower slope the volcanic basement is covered by a thick marine seismic sequence, characterized by channel-levee complexes. Two seismic sequences, respectively named Barano 1 (lower unit) and Barano 2 (upper unit) have been identified in the volcanic substratum and tentatively correlated with the Lower and Upper Formations of the Scarrupata di Barano.

The Punta San Pancrazio offshore is characterized by similar geological characteristics with respect to the Barano offshore. The seismic sequences Barano 1 and Barano 2 have been identified in the volcanic acoustic basement. A lava dome has been identified at the top of the Barano 1 sequence, laterally grading to a palaeocanyon filled by an acoustically-transparent seismic unit, probably pyroclastic in nature. The slope out of S. Pancrazio offshore is characterized by a marine turbidite sequence, overlying a volcanic acoustic basement and characterized by parallel to prograding seismic reflectors. The San Pancrazio promontory is composed by a sequence of lavas pertaining to the Monte di Vezzi volcanic complex, interstratified with pyroclastic deposits. The upper part of this sequence includes the Green Tuffs of Epomeo (Orsi et al., 2004).

Further information on the seismo-stratigraphic setting of the Ischia-Procida offshore are based on the interpretation of the L51 seismic profile. The line shows a buried volcanic edifice overlain by a prograding wedge and by the volcanites and the volcaniclastites of the Magnaghi canyon. The aggrading and prograding sequences of the continental shelf of Procida-Ischia have been identified. The stratigraphic relationships with the volcanic structure of the Ischia Bank are shown by the seismic section L57.

The contribution of the seismic stratigraphy to the knowledge of the geologic history of Ischia has allowed the implementation of the volcanologic and geologic aspects based on the outcrop data coupled to the new data of the seismic stratigraphy, as resulting from the new geologic map of Ischia at the scale 1:25.000 edited by the Region Campania, together with the ISPRA. The results here presented derive from the technical and scientific activities linked to the marine geological mapping of the Ischia island (geologic map n. 464 " Isola d'Ischia", scale 1:25.000 and 1:10.000).

Sorrento Peninsula – Capri Island

The geological setting of the sea bottoms of the Sorrento Peninsula-Capri island offshore is mostly influenced by the occurrence, in the onshore sector, of a Meso-Cenozoic structural high composed of shallow water carbonates pertaining to the marginal facies of the Campania-Lucania carbonate platform (D'Argenio et al., 1973). The Capri island is made of Mesozoic-Tertiary calcareous rocks and subordinately of sandstones and breccias. Its morphology is very similar to that of the Sorrento Peninsula, being Capri a south-westward extension of this structure to which, at Quaternary lower sea levels, it was attached (D'Argenio et al., 2012). The Capri island has a typical relief with two peaks at its NE and SW ends, sculptured in Mesozoic carbonate rocks and a central depression, partly covered by gravel tongues. The coastal cliffs are surrounded by a talus of calcareous breccia rims, with frequent small inlets with pocket beaches, often extending underwater. High pillar like blocks (faraglioni) that characterized the Capri coast are frequent along its south-eastern side. Down to 30 m of water depth several abrasion terraces may be observed, due to Quaternary lowering of sea level. The sea bottom morphology in the saddle between Capri and Sorrento is very shallow and shows morphological evidences of emersion (Quaternary sea lowering). The oldest rocks of Capri are Mesozoic in age and crop out in its south-western sector. They are made of different stratigraphic intervals often tectonically juxtaposed. Lower Jurassic dolomites and basinal cherty limestones and Middle Jurassic oolitic limestones are present. Other Mesozoic carbonate rocks crop out in the western sector of the island. They are constituted of Ellipsactinia limestones, Late Jurassic-Early Cretaceous in age. The remnant part of the Cretaceous is represented by pelagic cherty limestones (Scaglia calcarea), pink to white, showing in their lower part resedimented intercalations of calcarenites. In the northern sector of the island, almost at sea level, cherty limestones, Eocene-Oligocene in age, crop out, just below a thrust front that characterizes this sector of Capri. As to other Tertiary deposits, the synorogenic sandstones cropping out at the Bagni di Tiberio must also be quoted. The Quaternary deposits are mostly represented by coarse-grained clastic accumulations of breccias, colluvium, marine and small outcrops of pyroclastic deposits.

In the Sorrento Peninsula (Punta Campanella) the Mesozoic rocks are made of Cretaceous limestones with Tertiary calcarenites and sandstones. The underwater deposits have been surveyed in the first 30 metres also by direct observations and down to 200 metres by geophysical methodologies and sample dredging and coring. The area may be divided into several intervals normally of Holocene age, with a few outcrops of Pleistocene and Mesozoic not covered by recent sediments.

The following deposits, pertaining to different depositional environments, may be singled out from shallower to deeper areas (D'Argenio et al., 2012). The cliff talus deposits (Late Holocene) are composed of massive deposits of cobbles and boulders, made of angular to poorly rounded grey limestones and dolostones and carbonate light grey breccias and gravels. The submerged beach deposits (Late Holocene) are composed of fine grained gravels with sandy matrix, sometimes with well rounded to angular carbonate blocks, locally mixed with bioclasts. The inner shelf deposits (Holocene) are composed of sandy and silty deposits made of biolithoclastic, fine to medium sands and fine to medium carbonate sands, in a muddy matrix with clasts consisting of rock fragments, bioclasts and volcanoclasts (pumices and scoriae). The outer shelf deposits, Holocene in age, are composed of muds and sandy muds with a variable percentage of fine grained sands, sometimes organogenic, pertaining to the coastal detrital assemblage, in veneers overlying a silty sand subbottom; also common are mud supported organogenic gravels and bioclastic sands, typical of the muddy coastal detrital assemblage and rare bioclastic muddy sands and fine gravels, originated in situ from the breakdown of coralgal mounds. The upper slope deposits, Holocene in age, are composed of silty deposits, with a variable percentage of bioclastic and volcaniclastic fine sands, fairly abundant along the canyon heads; locally, off the narrow shelf in the southern sector of the island, bioclastic sands and gravels in a muddy matrix are common.

To the above deposits, Holocenic in age, older sediments of Late Pleistocene may be found (D'Argenio et al., 2012). They pertain to the following associations. Wedge shaped palimpsest deposits (muddy sands with Mollusk fragments and volcaniclastic deposits), along the shelf edge, covered by a thin veener of Holocene muds and palimpsest massive deposits, made of breccias and coarse gravels, covered by a thin blanket of Holocene muds and chaotic deposits, forming prominent and irregular relieves to the seabed (Lowstand System Tract deposits; D'Argenio et al., 2012).

The Island of Capri is located at the top of cliffy homoclinal, plunging to the NW and tectonically belongs to the Sorrento Peninsula structure, that separates the Gulf of Naples from the Gulf of Salerno. The rocks forming this linear element have a complex tectonic history documented by pre-Miocene synsedimentary deformation and by compressional (thrust and strike slip) features. The last uplift has left on the Capri island its signature as an extensional event to which the island has partly escaped, but of which it is also the final evidence (D'Argenio et al., 2012).

References

Acocella, V. & Funiciello, R. (1999). The interaction between regional and local tectonics during resurgent doming: the case of island of Ischia, Italy. *Journal of Volcanology and Geothermal Research*, *88*, 109-123.

Acocella, V., Funiciello, R. & Lombardi S. (1997). Active tectonics and resurgence at Ischia island (Southern Italy). *Il Quaternario*, *10*, 427-432.

Acocella, V., Funiciello, R., Marotta, E., Orsi G. & De Vita S. (2004). The role of extensional structures on experimental calderas and resurgence. *Journal of Volcanology and Geothermal Research*, *129*, 199-217.

Aiello, G., Budillon, F., de Alteriis, G., Di Razza, O., De Lauro, M., Marsella, E., Pelosi, N., Pepe, F., Sacchi, M. & Tonielli, R. (1997a). Seismic exploration of the perityrrhenian basins in the Latium-Campania offshore. Abstract International Congress ILP Task Force Origin of Sedimentary Basins, Torre Normanna (Palermo, Italy), June 1997.

Aiello, G., Budillon, F., de Alteriis, G., Ferranti, L., Marsella, E., Pappone, G., & Sacchi, M. (1997b). Late Neogene tectonics and basin evolution of the Southern Italy Tyrrhenian margin. Abstract International Congress ILP Task Force Origin of Sedimentary Basins, Torre Normanna (Palermo, Italy), June 1997.

Aiello, G., Budillon, F., Cristofalo, G., D'Argenio, B., de Alteriis, G., Ferraro, L., Marsella, E., Pelosi, N., Sacchi, M. & Tonielli, R. (2001). Marine geology and morphobathymetry in the Bay of Naples. In: Faranda F.M., Guglielmo L., Spezie G. (Eds.) Structures and Processes of the Mediterranean Ecosystems. Chapter 1, Springer-Verlag Italy, 1-8.

Aiello, G., Angelino, A., Marsella, E., Ruggieri, S. & Siniscalchi, A. (2004). Carta magnetica di alta risoluzione del Golfo di Napoli (Tirreno meridionale). *Bollettino della Società Geologica Italiana*, *123*, 333-342.

Aiello, G., Angelino, A., D'Argenio, B., Marsella, E., Pelosi, N., Ruggieri, S., & Siniscalchi, A. (2005). Buried volcanic structures in the Gulf of Naples (Southern Tyrrhenian sea, Italy) resulting from high resolution magnetic survey and seismic profiling. *Annals of Geophysics*, *48*, 1-15.

Aiello, G., Marsella, E. & Passaro, S. (2009a). Submarine instability processes on the continental slopes off the Campania Region (Southern Tyrrhenian sea, Italy): the case history of Ischia island (Naples Bay). *Bollettino di Geofisica Teorica Applicata*, *50* (2), 193-207.

Aiello, G., Budillon, F., Conforti, A., D'Argenio, B., Putignano, M. L., & Toccaceli, R. M. (2009b). Progetto per la realizzazione di cartografia geologica marina secondo le modalità CARG. Note illustrative alla cartografia geologica marina. Foglio geologico n. 464 "Ischia". Preprints, 25 May 2009, Regione Campania, Settore Geotecnica, Geotermia e Difesa Suolo, III SAL Geologia Marina, IAMC, Istituto per l'Ambiente Marino Costiero, Geomare Sud, CNR, Napoli.

Aiello, G., Marsella, E., Di Fiore, V. & D'Isanto, C. (2009c). Stratigraphic and structural styles of half-graben offshore basins in Southern Italy: multichannel seismic and Multibeam morpho-bathymetric evidences on the Salerno Valley (Southern Campania continental margin, Italy). *Quaderni di Geofisica*, *77*, 1-33.

Aiello, G., Marsella, E. & Ruggieri S. (2010). Three-dimensional magneto-seismic reconstruction of the Torre del Greco submerged volcanic structure (Naples Bay, Southern Tyrrhenian sea, Italy): implications for the Vesuvius's marine geophysics and volcanology. *Near Surface Geophysics*, *8* (1), 17-31.

Aiello, G., Cicchella, A. G., Di Fiore, V. & Marsella, E. (2011a). New seismostratigraphic data of the Volturno Basin (northern Campania, Tyrrhenian margin, southern Italy): implications for tectono-stratigraphy of Campania and Latium sedimentary basins. *Annals of Geophysics*, *54* (3), 265-283.

Aiello, G., Marsella, E., Cicchella A. G. & Di Fiore, V. (2011b). New insights on morpho-structures and seismic stratigraphy on the Campania continental margin based on deep multichannel seismic profiles. *Rendiconti Lincei Scienze Fisiche e Naturali*, *22* (4), 349-373.

Aiello, G., Marsella, E. & Passaro, S. (2012a). Stratigraphic and structural setting of the Ischia volcanic complex (Naples Bay, southern Italy) revealed by submarine seismic reflection data. *Rendiconti Lincei Scienze Fisiche e Naturali*, *23*, 387-408.

Aiello, G., Giordano L., Marsella, E. & Passaro, S. (2012b). Seismic stratigraphy and marine magnetics of the Naples Bay (southern Tyrrhenian sea, Italy): the onset of new technologies in marine data acquisition, processing and interpretation. In: Stratigraphic Analysis of Layered Deposits (Elitok O., Ed.), Chapter 2, 21-60.

Aiello, G., Marsella, E. & Di Fiore, V. (2012c). New seismostratigraphic and marine magnetic data of the Gulf of Pozzuoli (Naples Bay, Tyrrhenian sea, Italy): inferences for the tectonic and magmatic events of the Phlegrean Fields volcanic complex (Campania). *Marine Geophysical Research*, *33* (2), 97-125.

Alberico I., Di Fiore, V., Rapolla, A., Piemontese, L., Marsella, E., Giordano, L. & Iavarone, R. (2013). La valutazione della pericolosità da tsunami a scala locale, il caso studio della città di Napoli. *Miscellanea INGV*, 19, Riassunti Congresso AIQUA 2013, Napoli, 19-21 giugno 2013.

Alessio, M., Allegri, L., Antonioli, F., Belluomini, G., Improta, S., Manfra, L. & Preite Martinez, M. (1996). La curva di risalita del mare Tirreno negli ultimi 43 ka ricavata da datazioni su speleotemi sommersi e dati archeologici. Atti Int. Meeting on Underwater Geology GEOSUB '94, 8-10 giugno 1994, Palinuro (SA), *Memorie Descrittive della Carta Geologica d'Italia*, *52*, 261-276.

Antonioli, F. & Ferranti, L. (1986). Evidenze geomorfologiche sommerse relative ad oscillazioni del livello del mare: l'esempio di Capri (Grotta Azzurra e Grotta di Tragara). Atti del XXVI Congresso Geografico Italiano, Genova, 4-9 maggio 1992, 341-352.

Balducci, A., Vaselli, M. & Verdiani, G. (1985). Exploration well in the Ottaviano permit, Italy. In: Strub A.S. and Ungemach P. (Eds.) European Geothermal Update, Proceedings 3rd International Seminar on the Results of the EC Geothermal Energy Research, Reidel, Dordrecht.

Baratta, M. (1897). Il Vesuvio e le sue eruzioni. Dall'anno 79 D.C. al 1896, Roma.

Baratta, M. (1901). I terremoti d'Italia. Fratelli Bocca, Torino.

Barattolo, F., Cinque, A., D'Alessandro, E., Guida, M., Romano, P. & Russo Ermolli, E. (1992). Geomorfologia ed evoluzione tettonica quaternaria dell'Isola di Capri. *Studi Geologici Camerti*, Vol. Spec. 1992, 221-229.

Barberi, F., Innocenti, F., Lirer, L., Munno, R., Pescatore, T. S. & Santacroce, R. (1978). The Campanian Ignimbrite: a major prehistoric eruption in the Neapolitain area (Italy). *Bulletin of Volcanology*, *41* (1), 10-27.

Barra, D., Cinque, A., Italiano, A. & Scorziello, R. (1992a). Il Pleistocene superiore marino di Ischia: paleoecologia e rapporti con l'evoluzione tettonica recente. Studi Geologici Camerti, *Supplemento*, *1*, 231-243.

Barra, D., Cinque, A., Italiano, A. & Scorziello, R. (1992b). La serie marina olocenica di Cafieri (isola d'Ischia): implicazioni vulcanotettoniche e geomorfologiche. *Il Quaternario*, *5* (1), 17-26.

Bartole, R. (1984). Tectonic structures of the Latian-Campanian shelf (Tyrrhenian sea). *Bollettino Oceanologia Teorica Applicata*, *2*, 197-230.

Bartole R., Savelli, D., Tramontana, M. & Wezel F. C. (1983). Structural and sedimentary features in the Tyrrhenian margin of Campania, Southern Italy. *Marine Geology*, *55*, 163-180.

Belousov, A., Belousova, M. & Voight, B. (1999). Multiple edifice failures, debris avalanches and associated eruptions in the Holocene history of Shiveluch volcano, Kamchatka, Russia. *Bulletin of Volcanology*, *61*, 324-342.

Bernabini, M., Latmiral, G., Mirabile, L. & Segre, A. (1973). Alcune prospezioni sismiche per riflessione nei Golfi di Napoli e Pozzuoli. CIESM, Munchen.

Bernasconi, A., Bruni, P., Gorla, L., Principe, C. & Sbrana, A. (1981). Risultati preliminari dell'esplorazione geotermica profonda nell'area vulcanica del Somma-Vesuvio. *Rendiconti della Società Geologica Italiana*, *4*, 237-240.

Bigi, G., Coli, M., Cosentino, D., Parotto, M., Praturlon, A., Sartori, R., Scandone, P. & Turco, E. (1992). Structural model of Italy. S.E.L.C.A., Firenze (1992), Sheet 4, scale 1:500.000.

Blockley, S. P. E., Ramsey, C. B. & Pyle, D. M. (2008). Improved age modelling and high-precision age 3 estimates of Late Quaternary tephras, for accurate paleoclimate reconstruction. *Journal of Volcanology and Geothermal Research*, *177*, 251-262.

Brown, R. J., Orsi, G. & De Vita S. (2008). New insights into Late Pleistocene explosive volcanic activity and caldera formation on Ischia (southern Italy). *Bulletin of Volcanology*, *70*, 583-603.

Bruno, P. P. G., Cippitelli, G. & Rapolla, A. (1998). Seismic study of the mesozoic carbonate basement around Mt. Somma-Vesuvius volcanic complex (Italy). *Journal of Volcanology and Geothermal Research*, *84*, 311-322.

Bruno, P. P. G., de Alteriis, G. & Florio, G. (2002). The western undersea section of the Ischia volcanic complex (Italy, Tyrrhenian sea) inferred from marine geophysical data. *Geophysical Research Letters*, *29* (9), 1343, doi:10.1029/2001GL013904.

Bruno, P. P. G., Rapolla, A. & Di Fiore, V. (2003). Structural settings of the Bay of Naples (Italy) by seismic reflection data: implications for Campanian volcanism. *Tectonophysics*, *372*, 193-213.

Budillon, F., Violante, C. & De Lauro, M. (2003a). I fondali delle isole flegree: geologia e morfologia. In: Gambi M.C., De Lauro M., Iannuzzi F. (Eds.) Ambiente marino costiero e territorio delle isole flegree (Ischia, Procida e Vivara – Golfo di Napoli): Risultati di uno studio multidisciplinare. Memorie Accademia Scienze Fisiche e Matematiche in Napoli, 3-25.

Budillon, F., de Alteriis, G., Ferraro, L., Marsella, E., Molisso, F., Tonielli, R. & Violante, C. (2003b). I fondali dell'isola d'Ischia: morfobatimetria e geologia tra Punta del Chiarito e Punta San Pancrazio. In: Gambi M.C., De Lauro M., Iannuzzi F. (Eds.) Ambiente marino costiero e territorio delle isole flegree (Ischia, Procida e Vivara – Golfo di Napoli): Risultati di uno studio multidisciplinare. Memorie Accademia Scienze Fisiche e Matematiche in Napoli, 83-102.

Bunz, S., Mienert, J. & Berndt, C. (2003). Geological controls on the Storegga gas-hydrate system of the mid-Norwegian continental margin. *Earth and Planetary Science Letters*, *209* (3-4), 291-307.

Canals, M., Lastras, G., Urgeles, R., Casamor, J. L., Mienert, J., Cattaneo, A., De Batist, M., Haflidason, H., Imbo, Y., Laberg, J. S., Locat, J., Long, D., Longva, O., Masson, D., Sultan, N., Trincardi, F. & Bryn, P. (2004). Slope failure dynamics and impacts from seafloor and shallow sub-seafloor geophysical data: case studies from the COSTA project. *Marine Geology*, *213*, 9-72.

Capotondi, L. , Asioli, A. & Borsetti, A. M. (1994). Ecozonazione a foraminiferi plantonici e paleooceanografia dell'ultima deglaciazione del Mediterraneo centrale. *Atti X Congresso AIOL Alassio*, 4-6 Novembre 1992, 107-115.

Cartwright, J., Frey-Martinez, J. & James, D. (2005). Frontal confined versus frontally emerged submarine landslides: a 3D seismic characterization. Second International Conference on Submarine Mass Movements and Their Consequences, Oslo, Norway, 5-7 September, 24-25.

Cassano, E. & La Torre, P. (1987). Geophysics. In: Santacroce R. (Ed.) Somma-Vesuvius. Quaderni De La Ricerca Scientifica, CNR, Roma, Italy.

Catalano, R. et al. (1996). Linee guida alla cartografia geologica marina, scala 1:50.000, Gruppo di Lavoro del Servizio Geologico Nazionale, Bozza n. 1.

Chappell, J. & Shackleton, N. J. (1986). Oxygen isotopes and sea level. *Nature*, 324, 137-140.

Chiesa, S., Cornette, Y., Forcella, F., Gillot, P. Y., Pasquarè, G. & Vezzoli, L. (1985). Carta geologica dell'isola d'Ischia. CNR, Progetto Finalizzato Geodinamica, Roma, Italy.

Chiocci, F. L., Aiello, G., Baraza J., De Lauro, M., Ercilla, G., Estrada, F., Tonielli, R., Farran, M., Martorelli, E., Senatore, M. R., Tommasi, P., Bosman, A., Cristofalo, G., Ortolani, U. & Sarli, R. (1998a). Ischia debris avalanche: first evidence of a lateral collapse of the Ischia southern flank from TOBI images. Proceedings International Conference on Hard Soils and Soft Rocks, Naples, Italy, October 1998.

Chiocci, F. L., de Alteriis, G. (2006). The Ischia debris avalanche: first clear submarine evidence in the Mediterranean of a volcanic island prehistorical collapse. *Terra Nova*, *18*, 202-209.

Chiocci, F. L. & Casalbore, D. (2011). Submarine gullies on Italian upper slopes and their relationship with volcanic activity revisited 20 years after Bill Normark's pioneering work. *Geosphere*, *7*, 1284-1293.

Cinque, A. (1986). Guida alle escursioni geomorfologiche: Penisola Sorrentina, Capri, Piana del Sele e Monti Picentini. Dipartimento di Scienze della Terra, *pubblicazione*, *33*, 49-55.

Cinque, A. & Putignano, M. L. (1992). Geomorphology of the continental shelf around the Sorrento Peninsula (Southern Italy). *Giornale di Geologia*, *54* (2), 115-125.

Cinque, A. & Romano, P. (1990). Segnalazione di nuove evidenze di antiche linee di riva in Penisola Sorrentina (Campania). *Geografia Fisica Dinamica Quaternaria*, *13* (1), 23-36.

Cinque, A., Rolandi, G. & Zamparelli, V. (1985). L'estensione dei depositi marini olocenici nei Campi Flegrei in relazione alla vulcano-tettonica. *Bollettino della Società Geologica Italiana*, *104*, 327-348.

Cinque, A., Aucelli, P. P. C., Brancaccio, L., Mele, R., Milia, A., Robustelli, G., Romano, P., Russo, F., Santangelo, N. & Sgambati, D. (1997). Volcanism, tectonics and recent geomorphological change in the Bay of Napoli. *Supplementi Geografia Fisica Dinamica Quaternaria*, *III*, 123-141.

Civetta, L., Gallo, G. & Orsi, G. (1991). Sr and Nd isotope and trace element constraints on the chemical evolution of the magmatic system of Ischia (Italy) in the last 55 ky. *Journal of Volcanology and Geothermal Research*, *46*, 213-230.

Civetta, L., Orsi, G., Pappalardo, L., Fisher, R.V., Heiken, G. & Ort, M. (1997). Geochemical zoning, mingling, eruptive dynamics and depositional processes – the Campanian Ignimbrite, Campi Flegrei caldera, Italy. *Journal of Volcanology and Geothermal Research*, *75* (3-4), 183-219.

Coleman, J. M., Prior, D. B. & Lindsay, J. F. (1983). Deltaic influence on shelf edge instability processes. In: The Shelf Break: Critical Interface on Continental Margins (Stanley D.J. and Moore G.T., Eds.). *SEPM Special Publication*, *33*, 121-137.

D'Ambra, N. (1992). Ischia Perduta. Le grotte di Mavone alla Scannella. Centro di Ricerche Storiche D'Ambra, Forio d'Ischia, 1992/93, 10, ill. (Estr.).

D'Antonio, M., Tonarini, S., Arienzo, I., Civetta, L. & Di Renzo, V. (2007). Components and processes in the magma genesis of the Phlegrean Volcanic District, Southern Italy. In: Beccaluva L., Bianchini G., Wilson M. (Eds.) Cenozoic volcanism in the Mediterranean area. *Geological Society of America Special Paper*, *418*, 203-220.

D'Argenio, B. (1976). Le piattaforme carbonatiche peri-adriatiche. Una rassegna di problemi nel quadro geodinamico mesozoico dell'area periadriatica. *Memorie della Società Geologica Italiana*, *28*, 1-28.

D'Argenio, B., Pescatore, T. & Scandone, P. (1973). Schema geologico dell'Appennino meridionale (Campania e Lucania). *Atti Accademia Nazionale dei Lincei, Convegno "Moderne vedute sulla Geologia dell'Appennino"*, *183*, 49-72.

D'Argenio, B., Aiello, G., et al. (2004). Digital Elevation Model of the Naples Bay and adjacent areas, Eastern Tyrrhenian sea. In: Atlante di Cartografia Geologica scala 1:50.000 (progetto CARG), Servizio Geologico d'Italia (APAT), 32nd International Congress "Firenze 2004", Editore De Agostini, Italy.

D'Argenio, B., Violante, C., Sacchi, M., Budillon, F., Pappone, G., Casciello, E. & Cesarano, M. (2004). Carta Geologica di Capri, Bocca Piccola e Punta Campanella. In G. Pasquarè, C. Venturini, G. Groppelli (Eds.) Mapping Geology in Italy. APAT, 35-42.

D'Argenio, B., Barattolo, F., Budillon, F., Cesarano, M., Donadio, C., Pappone, G., Pugliese, A. & Putignano, M. L. (2012). Note illustrative della carta geologica alla scala 1:10.000, Foglio 484 Isola di Capri. Regione Campania, Assessorato Difesa del Suolo, Systemcart, Roma, Italy, 106.

Day, S. J. (1996). Hydrothermal pore fluid pressure and the stability of porous, permeable volcanoes. In: Mc. Guire W.J., Jones W.J. and Neuberg J. (Eds.) Volcano instability on the Earth and other planets. *Geological Society of London Special Publication*, *110*, 77-93.

de Alteriis, G. & Toscano, F. (2003). Introduzione alla geologia dei mari circostanti le isole flegree di Ischia, Procida e Vivara. In: Gambi M.C., De Lauro M., Iannuzzi F. (Eds.) Ambiente Marino Costiero e Territorio delle Isole Flegree (Ischia, Procida e Vivara – Golfo di Napoli): risultati di uno studio multidisciplinare. *Mem. Acc. Sc. Fis. e Mat. in Napoli*, *5*, 3-25.

de Alteriis, G. & Violante, C. (2009). Catastrophic landslides off Ischia volcanic island (Italy) during prehistory. In: Violante C. (Ed.) *Geohazard in rocky coastal areas*, vol. *322*. Special Publication of Geological Society of London, 73-104.

de Alteriis, G., Tonielli, R., Passaro, S. & De Lauro, M. (2006). Isole flegree: Ischia e Procida. Batimetrie dei fondali marini della Campania. Scala 1:30.000. Liguori Editore, Napoli.

de Alteriis, G., Insinga, D. D., Morabito, S., Morra, V., Chiocci, F. L., Terrasi, F., Lubritto, C., Di Benedetto, C. & Pazzanese, M. (2010). Age of submarine debris avalanches and tephrostratigraphy offshore Ischia Island, Tyrrhenian sea, Italy. *Marine Geology*, *278*, 1-18.

Deino, A. L., Orsi, G., De Vita, S. & Piochi, M. (2004). The age of Neapolitan Yellow Tuff caldera-forming eruption (Campi Flegrei caldera, Italy). assessed by 40Ar/39Ar dating method. *Journal of Volcanology and Geothermal Research*, *133*, 157-170.

Denlinger, R. P. & Iverson, R. M. (2004). Granular avalanches across irregular three-dimensional terrain: 1. Theory and Computation. *Journal of Geophysical Research*, *109*, F01014, doi:10.1029/2003JF000085.

De Pippo, T., Pennetta, M., Terlizzi, F. & Valente, A (2007). Principali tipi di falesia nella Penisola Sorrentina e nell'isola di Capri: caratteri e lineamenti morfoevolutivi. *Bollettino della Società Geologica Italiana*, *126* (2), 181-189.

De Vita, P., Agrello, D. & Ambrosino, F. (2006). Landslide susceptibility assessment in ash-fall pyroclastic deposits surrounding Somma-Vesuvius: application of geophysical surveys for soil thickness mapping. Journal of Applied Geophysics, 59, 126-139, doi:10.1016/j.japgeo.20.05.09.001.

De Vita, P., Celico, P., Di Clemente, E. & Rolandi, P. (2007). Engineering geological models of the initial landslides occurred on 30 April 2006, at Mount di Vezzi (Ischia island). *Italian Journal of Engineering Geology and Environment*, *2*.

Di Fiore, V., Aiello, G. & D'Argenio, B. (2011). Gravity instabilities in the Dohrn canyon (Bay of Naples, Southern Tyrrhenian sea): potential wave and run-up (tsunami) reconstruction from a fossil submarine landslide. *Geologica Carpathica*, *62* (1), 55-63.

Di Girolamo, P. & Rolandi, G. (1976). Vulcanismo sottomarino latite-basaltico latitico nel canale d'Ischia. *Rendiconti Accademia Scienze Matematiche e Fisiche in Napoli*, s.4-42, 1-36.

Di Girolamo, P., Ghiara, M. R., Lirer, L., Munno, R., Rolandi, G. & Stanzione, D. (1984). Vulcanologia e petrologia dei Campi Flegrei. *Bollettino della Società Geologica Italiana*, *103*, 349-370.

Di Maio, R., Piegari, E., Scotellaro, C. & Soldovieri, M. G. (2007). Tomografie di resistività per la definizione dello spessore e del contenuto d'acqua delle coperture piroclastiche a M.te di Vezzi (Isola d'Ischia) per lo studio degli eventi franosi dell'Aprile 2006. *Italian Journal Engineering Geology and Environment*, *2*, 65-72.

Di Vito, M. A., Isaia, R., Orsi, G., Southon, J., De Vita, S., D'Antonio, M., Pappalardo, L. & Piochi, M. (1999). Volcanism and deformation since 12.000 years at the Campi Flegrei caldera (Italy) *Journal of Volcanology and Geothermal Research*, *91*, 221-246.

Donadio, C., Romano, P. & Sgrosso, A. (1996). Ricerche di geomorfologia subacquea in Penisola Sorrentina. Atti International Meeting on Underwater Geology GEOSUB 94, 8-10 giugno 1994, Palinuro (SA), *Memorie Descrittive della Carta Geologica d'Italia*, *52*, 407-410.

Donadio, C. & Sgrosso, A. (2001). Elementi geologici e geomorfologici indicatori di rapidi mutamenti del paesaggio campano. Atti Convegno Internazionale "Politiche per la Tutela del Territorio", 23 marzo 2001, Napoli, 541-548.

Dvorak, J. J. & Mastrolorenzo, G. (1991). The mechanisms of recent vertical crustal movements in Campi Flegrei caldera., southern Italy. *Geological Society of America Special Paper*, *263*, 1-47.

Elsworth, D. & Day, S. J. (1999). Flank collapse triggered by intrusion: the Canarian and Cape Verde archipelago. *Journal of Volcanology and Geothermal Research*, *94*, 323-340.

Elverhoi, A., Norem, H., Andersen, E. S., Dowdeswell, J. A., Fossen, I., Haflidason, H., Kenyon, N. H., Laberg, J. S., King, E. L., Sejrup, H. P., Solheim, A. & Vorren, T. (1997). On the origin and flow behaviour of submarine slides on deep-sea fans along the Norwegian-Barents sea continental margin. *Geomarine Letters*, *17*, 119-125.

Embley, R. W. (1982). Anatomy of some Atlantic margin sediment slides and some comments on ages and mechanisms. In: Marine Slides and Other Mass Movements (Saxov S. and Niewenhuis J.K., Eds.), 189-213, Plenum Press, New York.

Erismann, T. H. & Abele, G. (2001). Dynamics of Rockslides and Rockfalls. Publications of Springer-Verlag, Berlin.

Fabbri, A., Gallignani, P. & Zitellini, N. (1981). *Geologic evolution of the peri-Tyrrhenian sedimentary basins of Mediterranean margins*. In: Wezel F.C. (Ed.) Sedimentary Basins of Mediterranean Margins: Bologna (Tecnoprint), 101-126.

Fabbri, A., Argnani, A., Bortoluzzi, G., Correggiari, A., Gamberi, F., Ligi, M., Marani, M., Penitenti, D., Roveri, M. & Trincardi, F. (2002). Carta geologica dei mari italiani alla scala 1:250.000. Guida al rilevamento. Presidenza del Consiglio dei Ministri, Dipartimento per i Servizi Tecnici Nazionali, Servizio Geologico, *Quaderni serie III*, volume *8*.

Fedele, F. G., Giaccio, B., Isaia, R. & Orsi, G. (2002). The Campanian Ignimbrite eruption, Heinrich Event 4 and Paleolithic change in Europe: a high resolution investigation, *Geophysical Monograph*, *139*, 301-325.

Fedele, F. G., Giaccio, B. & Hajdas I. (2008). Timescales and cultural processes at 40.000 BP in the light of the Campanian Ignimbrite eruption, western Eurasia. *Journal Human Evolution*, *55* (5), 834-857.

Ferranti, L. & Antonioli, F. (2007). Misure del solco tirreniano (MIS 5.5). nell'Isola di Capri: implicazioni su micro-dislocazioni e blocchi cinematici attivi negli ultimi 124 ka. *Il Quaternario*, *20* (2), 125-136.

Finetti, I. & Morelli, C. (1973). Esplorazione sismica a riflessione dei Golfi di Napoli e Pozzuoli. *Bollettino di Geofisica Teorica Applicata*, *16*, 62-63, 175-222.

Forcella, F., Gnaccolini, M. & Vezzoli, L. (1983). I depositi piroclastici del settore sud-orientale dell'Isola d'Ischia. *Rivista Italiana Paleontologia Stratigrafia*, *89*, 1-35.

Francis, P. W. (1993). Volcanoes: a Planetary Perspective. Oxford University Press., 443.

Fischer, R. V., Glicken, H. X. & Hoblitt, R. P. (1987). Mount St. Helens deposits in South Coldwater Creek, Washington. *Journal of Geophysical Research*, *92*, 10267-10283.

Frey-Martinez, J., Cartwright, J. & James, D. (2006). Frontally confined versus frontally emerged submarine landslides: a 3D seismic characterization. *Marine and Petroleum Geology*, *23*, 585-604.

Fusi, N. (1996). Structural settings of the carbonatic basement in the Gulf of Naples and its relationship with magma uprising. *Annals of Geophysics*, *39* (3), 439-509.

Fusi, N., Mirabile, L., Camerlenghi, A. & Ranieri, G. (1991). Marine geophysical survey of the Gulf of Naples (Italy): relationship between volcanic activity and sedimentation. *Memorie della Società Geologica Italiana*, *47*, 95-114.

Galloway, W. E. (1998). Siliciclastic slope and base-of-slope depositional systems: component facies, stratigraphic architecture, and classification: *American Association of Petroleum Geologists Bulletin*, *82* (4), 569-595.

Gawthorpe, R. L. & Clemmey, H. (1985). Geometry of submarine slides on the Barra Fan, west of Britain. *Geological Society of London Special Publications*, *129*, 67-79.

Gillot, P. Y., Chiesa, S., Pasquarè, G. & Vezzoli, L. (1982). 33.000 yr K/Ar dating of the volcano-tectonic horst of the isle of Ischia, Gulf of Naples. *Nature*, *229*, 242-245.

Goodwin, R. H. & Prior, D. B. (1989). Geometry and depositional sequences of the Mississippi Canyon, Gulf of Mexico. *Journal of Sedimentary Petrology*, *59* (2), 318-329.

Guadagno, F. M. & Mele, R. (1994). Evoluzione da frana in aree campane di interesse ambientale. In: Territorio ed Ambiente: Giornate Poster sulle ricerche del Gruppo geomineralogico, *IV Conferenza Scientifica Annuale sulle Attività di Ricerca dei Dipartimenti*, 21-23 aprile 1994, Napoli, 134-137.

Guadagno, F. M. & Mele, R. (1995). La fragile isola d'Ischia. *Geologia Applicata e Idrogeologia*, *30* (1), 177-187.

Hampton, M. A., Lee, H. J. & Locat, J. (1996). Submarine landslides. *Review of Geophysics*, *34* (1), 33-59.

Heinrich, P., Boudon, G., Komarowski, J. C., Sparks, R. S. J., Herd, R. & Voight, B. (2001). Numerical simulation of the December 1997 debris avalanche in Montserrat, Lesser Antilles. *Geophysical Research Letters*, *28*, 2529-2532.

Helland-Hansen, W. & Gjielberg, J. G. (1994). Conceptual basis and variability in sequence stratigraphy: a different perspective. *Sedimentary Geology*, *92*, 31-52.

Holmes, R., Long, D. & Dodd, L. R. (1998). Large-scale debrites and submarine landslides on the Barra Fan, west of Britain. *Geological Society of London Special Publications*, *129*, 67-79.

Kastens, K., Mascle, J., Auroux, C., Bonatti, E., Broglia, C., Channel, J. E. T., Curzi, P., Emeis, K. C., Glason, G., Hasegawa, S., Hieke, W., Mascle, G., McCoy, F., McKenzie, J., Mendelson, J., Mueller, C., Rehault, J. P., Robertson, A., Sartori, R., Sprovieri, R. & Torii, M. (1986). Young Tyrrhenian sea evolved very quickly. *Geotimes*, *31* (8), 11-14.

Kastens, K., Mascle, J., Auroux, C., Bonatti, E., Broglia, C., Channel, J. E. T., Curzi, P., Emeis, K. C., Glacon, C., Asegawa, S., Hieke, W., Mascle, G., McCoy, F., McKenzie, J., Mendelson, G., Muller, C., Rehault, J. P.,

Robertson, A., Sartori, R., Sprovieri, R. & Torii, M. (1988). ODP Leg 107 in the Tyrrhenian sea: insights into passive margin and back-arc basin evolution. *GSA Bulletin*, *100*, 1140-1156.

King, E. L., Sejrup, H. P., Haflidason, H., Elverhoi, A. & Aarseth, I. (1996). Quaternary seismic stratigraphy of the North Sea Fan: glacially fed gravity flow aprons, hemipelagic sediments and large submarine slides. *Marine Geology*, *130*, 293-315.

Insinga, D., Di Meglio, A., Molisso, F. & Sacchi, M. (2002). Stratigrafia e caratteristiche fisiche dei depositi olocenici del Porto di Miseno, Golfo di Pozzuoli (Tirreno centro-orientale). *Il Quaternario*, *15* (2), 11-21.

Iovino, M. & Perriello Zampelli, S. (2007). Le frane del 30 Aprile 2006 al M.te di Vezzi (Ischia, NA) nel contesto delle frane di scivolamento valanga-colata nei suoli vulcanici in Campania. *Italian Journal of Engineering Geology and Environment*, *2*.

Ippolito, F., ,Ortolani F. & Russo, M. (1973). Struttura marginale tirrenica dell'Appennino Campano: reinterpretazione di dati di antiche ricerche di idrocarburi. *Memorie della Società Geologica Italiana*, *XII*, 227-250.

Judenherc, S. & Zollo, A. (2004). The Bay of Naples (Southern Italy): constraints on the volcanic structures inferred from a dense seismic survey. *Journal of Geophysical Research*, *109*, B10312, doi: 10.1029/ 2003JB002876.

Laberg, J. S. & Vorren, T. O. (2000). The Traenadjupet Slide, offshore Norway – morphology, evacuation and triggering mechanisms. *Marine Geology*, *171*, 95-114.

Latmiral, L., Segre, A. G., Bernabini, M. & Mirabile, L. (1971). Prospezioni sismiche per riflessione nei Golfi di Napoli e Pozzuoli ed alcuni risultati geologici. *Bollettino della Società Geologica Italiana*, *90*, 163-172.

Lipman, P. W. & Mullineaux, D. R. (Eds.) The 1980 Eruptions of Mount Sant Helens. United States Geological Survey Professional Paper, Washington, 1250.

Luongo, G., Cubellis, E. & Obrizzo, F. (1987). *Ischia: storia di un'isola vulcanica*. Liguori Editore, Napoli.

Kastens, K., Mascle, J., Auroux, C., Bonatti, E., Broglia, C., Channell, J. E. T, Curzi, P., Emeis, K. C., Glason, G., Hasegawa, S., Hieke, W., Mascle, G., Mc Coy, F., Mc Kenzie, J., Mendelson, J., Muller, C., Rehault, J. P., Robertson, A., Sartori, R., Sprovieri, R. & Torii, M. (1986). Young Tyrrhenian sea evolved very quickly. *Geotimes*, *31* (8), 11-14.

Kastens, K., Mascle, J., Auroux, C., Bonatti, E., Broglia, C., Channell, J. E. T, Curzi, P., Emeis, K. C., Glason, G., Hasegawa, S., Hieke, W., Mascle, G.,

Mc Coy, F., Mc Kenzie, J., Mendelson, J., Muller, C., Rehault, J. P., Robertson, A., Sartori, R., Sprovieri, R. & Torii, M. (1988). ODP Leg 107 in the Tyrrhenian sea: insights into passive margin and back-arc basin evolution. *Geological Society of America Bulletin*, *100*, 1140-1156.

Mallet, R. (1852-1854). Catalogue of recorded earthquakes 1606 B.C.-1850 A.D. Report on the state of science. Third report on the facts of earthquake phenomena. British Association for the Advancement of Science, London.

Mariani, M. & Prato, R. (1988). I bacini neogenici costieri del margine tirrenico:approccio sismico-stratigrafico. *Memorie della Società Geologica Italiana*, *41*, 519-531.

Marianelli P., Sbrana A., Proto M. (2006). Magma chamber of the Campi Flegrei supervolcano at the time of eruption of the Campanian Ignimbrite. *Geology*, *34* (11), 937-940.

McEwen, A. S. & Malin, M. C. (1989). Dynamics of Mount St. Helens's 1980. Pyroclastic flows, rockslide avalanches, lahars and blast. *Journal of Volcanology and Geothermal Research*, *37*, 205-231.

Mercalli, G. (1883). Vulcani e fenomeni vulcanici in Italia. Milano.

Mele, R. & Del Prete, S. (1998). Fenomeni di instabilità dei versanti in Tufo Verde del Monte Epomeo (Isola d'Ischia, Campania). *Bollettino della Società Geologica Italiana*, *117* (1), 93-112.

Micallef, A. & Mountjoy, J. J. (2011). A topographic signature of a hydrodynamic origin for submarine gullies, *Geology*, *39*, 115-118.

Milia A. (2000). The Dohrn canyon: a response to the eustatic fall and tectonic uplift of the outer shelf along the eastern Tyrrhenian sea margin, Italy. *Geomarine Letters*, *20* (2), 101-108.

Milia A. (2010). The stratigraphic sig*nature* of volcanism off Campi Flegrei (Bay of Naples, Italy). In: Groppelli G. and Goette L.V. (Eds.) Stratigraphy and geology of volcanic areas. GSA Special Papers, ISBN 9780813724645.

Milia, A., Mirabile, L., Torrente, M. M. & Dvorak, J. J. (1998). Volcanism offshore of Vesuvius volcano in the Naples Bay. *Bulletin of Volcanology*, *59* (6), 404-413.

Milia, A. & Torrente, M. M. (1999). Tectonics and stratigraphic architecture of a peri-tyrrhenian half-graben (Bay of Naples, Italy). *Tectonophysics*, *315* (1), 301-318.

Milia, A. & Torrente, M. M. (2007). The influence of paleogeographic setting and crustal subsidence on the architecture of ignimbrites in the Bay of Naples (Italy) *Earth and Planetary Science Letters*, *263*, 192-206.

Milia, A. & Torrente, M. M. (2012). A possible link between faulting, cryptodomes and lateral collapses at Vesuvius Volcano (Italy). *Global and Planetary Change*, 90-91, 121-134.

Milia, A., Molisso, F., Raspini, A., Sacchi, M. & Torrente, M. M. (2008). Syneruptive features and sedimentary processes associated with pyroclastic currents entering the sea: the AD 79 eruption of Vesuvius, Bay of Naples, Italy. *Journal of the Geological Society*, *165*, 839-848.

Minisini, D., Trincardi, F., Asioli, A., Canuz, M. & Foglini, F. (2007). Morphological variability of exposed mass-transport deposits on the eastern slope of Gela Basin (Sicily Channel). *Basin Research*, 19, 217-240.

Mitchum, J. R., Vail, P. R. & Sangree, J. B. (1977). Stratigraphic interpretation of seismic reflection pattern in depositional sequences. In: Payton C.E. (Ed.) Seismic stratigraphy – applications to hydrocarbon exploration. *AAPG Mem.*, *26*, 117-134.

Morhange, C., Marriner, N., Laborel, J., Todesco, M. & Oberlin, C. (2005). Rapid sea level movements and noneruptive crustal deformations in the Phlegrean Fields caldera, Italy. *Geology*, *34* (2), 93-96.

Mostardini, F. & Merlini, S. (1986). Appennino centro-meridionale. Sezioni geologiche e proposta di modello strutturale. *Memorie della Società Geologica Italiana*, *35*, 177-202.

Mulder, T. & Cochonat, P. (1996). Classification of offshore mass movements. *Journal of Sedimentary Research*, *66*, 43-57.

Normark, W. R., Mc Gann, M. & Slitter, R. (2004). Ages of Palos Verdes submarine debris avalanche, southern California. *Marine Geology*, *203*, 247-259.

Nunziata, C., Mele, R. & Natale, M. (1999). Shear wave velocities and primary influencing sectors of Campi Flegrei-Neapolitan deposits. *Engineering Geology*, *54*, 399-212.

O'Leary, D. W. & Dobson, M. R. (1992). Southeastern New England continental rise: Origin and history of slide complexes. In. Poag C.W. and De Graciansky P.G. (Eds.) Geological Evolution of Continental Rises, Van Nostrand Reinhold, New York, 214-265.

Orsi, G., Gallo, G. & Zanchi, A. (1991). Simple-shearing block-resurgence in caldera depressions: A model from Pantelleria and Ischia. *Journal of Volcanology and Geothermal Research*, *47*, 1-11.

Orsi, G., D'Antonio, M., De Vita, S. & Gallo, G. (1992). The Neapolitain Yellow Tuff, a large magnitudo trachytic phreatoplinian eruption: eruptive dynamics, magma withdrawal and caldera collapse. *Journal of Volcanology and Geothermal Research*, *53*, 275-287.

Orsi, G., Civetta, L., D'Antonio, M., Di Girolamo, P. & Piochi, M. (1995). Step-filling and development of a zoned magma chamber: the Neapolitain Yellow Tuff case history. *Journal of Volcanology and Geothermal Research*, *67*, 291-312.

Orsi, G., De Vita, S., Di Vito, M., Isaia, R., Nave, R. & Heiken, G. (2002). Facing volcanic and related hazard in the Neapolitain area. In: Heiken G., Fakundiny R., Sutter J. (Eds.) Earth Sciences in the city. American Geophysical Union Special Publication, Washington, 121-170.

Orsi, G., Di Vito, M. A. & Isaia, R. (2004). Volcanic hazard assessment at the restless Campi Flegrei caldera. *Bulletin of Volcanology*, *66*, 514-530.

Orsi, G., Di Vito, M. A., Selva, J. & Marzocchi, W. (2009). Long-term forecast of eruption style and size at Campi Flegrei caldera (Italy) Earth Planetary and Science Letters, *287*, 265-276.

Ortolani, F. & Aprile, F. (1978). Nuovi dati sulla struttura profonda della Piana Campana ad est del fiume Volturno. *Bollettino della Società Geologica Italiana*, *97*, 591-608.

Pabst, S., Worner, G., Civetta, L. & Tesoro. R. (2008). Magma chamber evolution prior to the Campania Ignimbrite and Neapolitan Yellow Tuff eruptions (Campi Flegrei, Italy). *Bulletin of Volcanology*, *70*, 961-976.

Pappalardo, L., Civetta, L., D'Antonio, M., Deino, A., Di Vito, M., Orsi, G., Carendente, A., De Vita, S. & Piochi, M. (1999). Chemical and Sr-isotopical evolution of the Phlegrean magmatic system before the Campanian Ignimbrite and the Neapolitan Yellow Tuff eruptions. *Journal of Volcanology and Geothermal Research*, *91* (2-4), 141-166.

Pasaric, M., Brizuela, B., Graziani, L., Maramai, A. & Orlic, M. (2012). Historical tsunamis in the Adriatic sea. *Natural Hazards*, *61* (2), 281-316.

Passaro, S. (2005). Integrazione di dati magnetici e morfobatimetrici in aree marine vulcaniche e non vulcaniche. Tesi di Dottorato di Ricerca in Scienze ed Ingegneria del Mare, Università degli Studi di Napoli Federico II, Napoli, Italy.

Pellegrino, A. (1994). I fenomeni franosi dell'area metropolitana napoletana. In: Rischi naturali ed impatto antropico nell'area metropolitana napoletana, a cura del CIRAM, Università degli Studi di Napoli "Federico II", Acta Neapolitana, Guida Editore, 237-256, Napoli.

Pescatore, T., Diplomatico, G., Senatore, M. R., Tramutoli, M. & Mirabile, L. (1984). Contributi allo studio del Golfo di Pozzuoli: aspetti stratigrafici e strutturali. *Memorie della Società Geologica Italiana*, *27*, 133-149.

Piper, D. J. W., Cochonat, P. & Morrison, M. L. (1999). The sequence of events around the epicentre of the 1929 Grand Banks earthquake: initiation of debris flows and turbidity current inferred from sidescan sonar. *Sedimentology*, *46*, 79-97.

Poli, S., Chiesa, S., Gillot, P.Y., Gregnanin, A. & Guichard, F. (1987) Chemistry versus time in the volcanic complex of Ischia (Gulf of Naples, Italy): evidence of successive magmatic cycles. *Contributions to Mineralogy and Petrology, 95 (3),* 322-335.

Posamentier, H. W. & Allen, H. W. (1993). Variability of the sequence stratigraphic model: effects of local basin factors. *Sedimentary Geology*, *86*, 91-109.

Posamentier, H.W. & Kolla, D. (2003). Seismic geomorphology and stratigraphy of depositional elements in deep-water settings. *Journal of Sedimentary Research*, *73*, 367-388.

Posamentier, H. W., Erskine, R. D. & Mitchum, R. M. (1991). Models for submarine fan deposition within a sequence stratigraphic framework. In: Weimer P. and Link M.H. (Eds.) Seismic Facies and Sedimentary Processes of Submarine Fans and Turbidite Systems, New York, Springer-Verlag, 127-136.

Pyle, D. M., Ricketts, G. D., Margari, V., van Andel, T. H., Sinitsyn, A. A., Praslov, N. D. & Lisitsyn, S. (2006). Wide dispersal and deposition of distal tephra during the Pleistocene Campanian Ignimbrite/Y5 eruption, Italy. *Quaternary Science Reviews*, *25* (21/22), 2713-2728.

Rolandi, G., Bellucci, F., Heizler, M. T., Belkin, E. & De Vivo, B. (2003). Tectonic controls on the genesis of ignimbrites from the Campanian Volcanic Zone, Southern Italy. *Mineralogy and Petrology*, *79* (1-2), 3-31.

Romano, P. & Sgrosso, I. (1992). L'evoluzione geomorfologica della grotta subacquea dell'Isca (Penisola Sorrentina, Campania). I Convegno Geologia Subacquea Sottomarina GEOSUB, 5-6 novembre 1991, Roma, *Giornale di Geologia di Bologna*, *54* (2), 151-163.

Rosi, M. & Sbrana, A. (1987). Phlegrean Fields. CNR, Quaderni De La Ricerca Scientifica, Roma, Italy.

Rosi, M., Vezzoli, L., Aleotti, P. & De Censi, M. (1996). Interaction between caldera collapse and eruptive dynamics during the Campanian Ignimbrite eruption, Phlegrean Fields, Italy. *Bulletin of Volcanology*, *57* (7), 541-554.

Rosi, M., Vezzoli, L., Castelmenzano, A. & Greco, G. (1999). Plinian pumice fall deposits of the Campanian Ignimbrite eruption, Phlegrean Fields, Italy. *Journal of Volcanology and Geothermal Research*, *91* (2-4), 179-198.

Ross, W. C., Halliwell, B. A., May, J. A., Watts, D. E. & Syvitski, J. P. M. (1994). Slope readjustement: a new model for the development of submarine fans and aprons. *Geology*, *22*, 511-514.

Ruggieri S., Aiello, G. & Marsella, E. (2007). Integrated marine geophysical data interpretation of the Naples Bay continental slope. *Bollettino di Geofisica Teorica Applicata*, *48*, 1-24.

Sacchi, M., Infuso, S. & Marsella, E. (1994). Late Pliocene-Early Pleistocene compressional tectonics in offshore Campania (Eastern Tyrrhenian sea). *Bollettino di Geofisica Teorica Applicata*, *36*, 141-144.

Sacchi, M., D'Argenio, B., Morra, V., Petrazzuoli, S., Aiello, G., Budillon, F., Sarnacchiaro, G. & Tonielli, R. (2000). Pyroclastic lumps: quick diapiric structures off the Naples Bay, Italy. *Geophysical Research Abstracts*, 2.

Santacroce, R. (1987). Somma Vesuvius. CNR, Quaderni De La Ricerca Scientifica, Roma, Italy.

Saxov, S. & Niewenhuis, J. K. (1982). Marine slides and other mass movements. Plenum Press, New York.

Scarpati, C., Cole, P. & Perrotta, A. (1993). The Neapolitain Yellow Tuff – a large volume multiphase eruption from Campi Flegrei, Southern Italy. *Bulletin of Volcanology*, *55*, 343-356.

Schiattarella, M. (1990). Random comparison of fracture patterns from Phlegrean Fields and carbonate horst of Sorrento Peninsula – Capri island, southern Italy. Relazioni scuola avanzata di Vulcanologia, Osservatorio Vesuviano, Ercolano.

Schwab, W. C., Lee, H. J. & Twichell, D. C. (1993). Submarine landslides: selected studies in the U.S. exclusive economic zone. Bull. 2002, U.S. Geological Survey, U.S., Dept. of Interior, Washington, D.C.

Secomandi, M., Paoletti, V., Aiello, G., Fedi, M., Marsella, E., Ruggieri, S., D'Argenio, B. & Rapolla, A. (2003). Analysis of the magnetic anomaly field of the volcanic district of the Bay of Naples, Italy. *Marine Geophysical Research*es, *24*, 207-221.

Sgrosso, I. (1986). Criteri ed elementi per una ricostruzione paleogeografica delle zone esterne dell'Appennino centro-meridionale. *Memorie della Società Geologica Italiana*, *35*, 203-228.

Signorelli, S., Vaggelli, G., Romano, C. & Carroll, M. (2001). Volatile element zonation in Campanian Ignimbrite magmas (Phlegrean Fields, Italy): evidence from the study of glass inclusions and matrix glasses. *Contributions to Mineralogy and Petrology*, *5*, 543-553.

Simini, M., La Torre, D. & Zollo, A. (2003). 3D limestone top morphology in Naples Bay. In: Capuano P., Gasparini A., Zollo A., Virieux J., Casale R.,

Yeroyanni M. (Eds.) The Internal Structure of Mount Vesuvius, 217-235, Liguori Editore, Napoli, Italy.

Sousa, J. & Voight, B. (1995). Multiple-pulses debris avalanche emplacement at Mounts St Helens in 1980: Evidence from numerical continuum flow simulations. *Journal of Volcanology and Geothermal Research*, *66*, 227-250.

Sultan, N., Cochonat, P., Canals, M., Cattaneo, A., Dennielou, B., Haflidason, H., Laberg, J. S., Long, D., Mienert, J., Trincardi, F., Urgeles, R., Vorren, T. & Wilson, C. (2004). Triggering mechanisms of slope instability processes and sediment failures on continental margins: a geotechnical approach. *Marine Geology*, *213*, 291-321.

Tappin, D. R. & Shipboard scientists (1999). Offshore surveys identify sediment slump as likely cause of devastating Papua New Guinea tsunami 1998. *Eos*, *80* (30), July 27, 329.

Tappin, D. R., Watts, P., Mc Murtry, G. M., Lafoy, Y. & Matsumoto, T. (2001). The Sissano, Papua New Guinea tsunami of July 1998 – Offshore evidence on the source mechanisms. *Marine Geology*, *175*, 1-23.

Tinti, S., Maramai, A. & Graziani, L. (2004). The new catalogue of Italian tsunamis. *Natural Hazards and Earth System Sciences*, *33* (3), 439-465.

Tinti, S., Chiocci, F. L., Zaniboni, F., Pagnoni, G. & de Alteriis, G. (2011). Numerical simulation of the tsunami generated by a past catastrophic landslide on the volcanic island of Ischia, Italy. *Marine Geophysical Researche*s, *32*, 287-297.

Trincardi, F. & Zitellini, N. (1987). The rifting of the Tyrrhenian basin. *Geomarine Letters*.

Trincardi, F., Correggiari, A. & Roveri, M. (1994). Late Quaternary transgressive erosion and deposition in a modern continental shelf: the Adriatic semienclosed basin. *Geomarine Letters*, *14*, 41-51.

Ui, T. (1983). Volcanic dry avalanche deposits – Identification and comparison with non-volcanic debris stream deposits. *Journal of Volcanology and Geothermal Research*, *18*, 135-160.

Vail, P. R., Mitchum, R. M., Todd, R. G., Widmier, J. M., Thompson, S., Sangree, J. B. & Bubb, J. N. (1977). Seismic stratigraphy and global change of sea level. In: Payton C.E. (Ed.) Seismic stratigraphy – Applications to Hydrocarbon Exploration. *AAPG Mem. 26*, 49-116.

Van der Meerwe, W. C., Hodgson, D. M. & Flint, S. T. (2011). Origin and terminal architecture of a submarine slide: a case study from the Permian Vischkuil Formation, Karoo Basin, South Africa. *Sedimentology*, *58* (7), 2012-2038.

Vezzoli, L. (1988). Island of Ischia. CNR, *Quaderni De La Ricerca Scientifica*, *114* (10), 122.

Vingiani, S. & Terribile, F. (2007). Suoli delle nicchie di distacco delle frane d'Ischia. *Italian Journal Engineering Geology and Environment*, *2*, doi:10.4408/IJEGE.2007-02.O-04.

Voight, B. (1983). Time scale for the first moments of the May 18 eruption. In: Lipman P.W. and Mullineaux D.R. The 1980 eruptions of Mounts St. Helens, Washington, vol. *1250*.

Wadge, G., Francis, P. W. & Ramirez C. F. (1995). The Socompa collapse and avalanche event. *Journal of Volcanology and Geothermal Research*, *66*, 309-366.

Walsh, E. O., Hacker, B. R., Gans, P. B., Groove, M. & Gehrels, G. (2007). Protolith ages and exhumation histories of ultrahigh-pressure rocks across the Western Gneiss Region, Norway. *GSA Bulletin*, *119*, 289-301.

Ward, S. N. & Day, S. (2006). Particulate kinematic simulation of debris avalanches: interpretation of deposits and landslide seismic signals of Mount Sant Helens, Washington, May 18. *Geophysical Journal International*, *167*, 991-1004.

Watts P. (2001). Some opportunities of the landslide tsunami hypothesis. *Sci Tsunami Hazards.*

Watts, P., Grilli, S. T., Tappin, D. R., Yalciner, A. C., Fryer, G. J. & Kroenke, L. W. (2001). Tsunami generation of submarine mass failure part II: case studies. Applied Fluids Engineering.

Wilson R. C. & Keefer, D. K. (1985). Predicting areal limits of earthquake-induced landsliding. Professional Paper 1360, U.S. Geological Survey, U.S. Dept. of Interior, Washington, D.C.

In: Continental Shelf
Editors: L. Scanlon and J. L. Ranieri
ISBN: 978-1-62948-330-6

Chapter 2

Geomorphological Characteristics of the Submerged Topography Along the Egyptian Red Sea Coast

Moawad Badawy*
Department of Biogeochemistry,
Max Planck Institute for Chemistry, Mainz, Germany,
Department of Geography, Faculty of Arts,
Ain Shams University, Cairo, Egypt

Abstract

A long time ago the Red Sea was only known by small-scale bathymetric, magnetic anomaly maps and a few seismic reflection or refraction profiles. Therefore, detection of the major submerged coastal features was unattainable. This study is based on the integration of different data sets of topography and bathymetry (e.g. the global bathymetry data set, the SRTM DTED-2, the soviet military topographic maps of scale 1:200.000 and the US army topographic maps of scale 1:250.000) to reveal the main submarine landforms that marked the continental shelf and its related slopes along the Egyptian Red Sea Coast from latitude 27°43`N to the Egyptian-Sudanese border at latitude 22°00`N. The study revealed the major types of coral reefs along the

* Phone: +49-(0) 6131 305 6502, Fax: +49-(0) 6131 305 6019, Email: moawad1974@ yahoo.com.

Egyptian Red Sea Coast. It is deduced that the continental shelf is noticeably influenced by the surface fault system extending eastward into the main Red Sea depression, showing the continental edge mostly like a fault-scarp of ~60° anticlockwise fault plane. Sea ridges and subbasins were distinguished at the lower toe of the continental slope, which seem to be a result of a regional fold system. Two sea peaks of extinct volcanoes were recognized. Two types of submarine canyons were recognized as deep incised Messinian canyons and shallow canyons. The deep incised canyons (~500 m bsl) carve the continental edge with remarkable steep walls. They might be formed as a result of the Messinian event (~5.59 Ma). The shallow canyons are mostly developed during the Pleistocene lower sea level (~90–130 m bsl) where the major wadis cut their water courses through the continental shelf. Some individual submerged deltas were identified, showing a close relationship with the present-day drainage system, although they were supposed to be produced by an ancestor drainage system. Notable submarine terraces were recognized at depths 20–25, 50–75, and 100–120 m bsl that are in agreement with the generalized global curve of sea-level rise since the LGM (~23–18 ka bp). It is expected that results of this study will be helpful for shedding light on the geomorphologic evolution of the continental shelf and its surrounding landmass.

1. Introduction

Depths of the Red Sea have been of great interest owing to the great controversy concerning structure and evolution of the Red Sea as the best example of a young ocean that evolved from an intracontinental rift to an axial oceanic trough (Guennoc et al., 1988). A long time ago, the northern Red Sea was only known by small-scale bathymetric and magnetic anomaly maps (Allan, 1970; Laughton, 1970; Hall, 1977) and by a few seismic reflection or refraction profiles (Drake and Girdler, 1964; Knott et al., 1966; Phillips and Ross, 1970; Tewfik and Ayyad, 1982; Guennoc et al., 1988). Therefore, neither the submerged coastal features nor the continental shelf has got serious attentions. Even the Admiralty charts of the Red Sea published in 1965 and their updates and/or reprinted versions show many sporadic points that are not convincing or adequate to ascertain the major submerged coastal topography. Another bathymetric map of scale 1:1000.000 was published by the International Hydrographic Bureau, Hydraulic Department-UK in 1970. The map was made of 500 m depth interval, showing many steep cliffs along the coastlines and somewhat close to the inshore. Marginal deeps were revealed as well on both sides of the northern Red Sea that may have considerable tectonic significances. The accuracy of the map was poor as no radio aids covered the area and very few soundings were controlled by satellite navigation (Laughton, 1970). Guennoc et al. (1988) established new bathymetric maps of scale 1:250.000 at latitude 26° N with a 100 m contour interval based on the base maps of Pautot et al. (1986) and Bäcker et al. (1975) and revised using regional sea-beam surveys. The maps were completed with 12 kHz or 3.5 Hz echo-sounder profiles done during the ''Geomerouad'' and ''Mines'' cruises. In 1999, the RlV Meteor Cruise M44/3 investigated some places in the northern Red Sea using a hydrosweep bathymetric multi-beam system and the narrow-beam parasound sediment echosounder (Pätzold, 2000). In 2002, the HMS Echo was launched at Appledore in Devon, England. It was designed to carry out a wide range of survey work, including support to submarine and amphibious operations, through the collection of oceanographic and bathymetric (analysis of the ocean, its salinity and sound profile) data. It left Devonport in the first week of 2011 on a two-year deployment to the Red Sea, the Gulf, the Indian Ocean, the Middle and Far East (U.K. Royal Navay, 2013). During its enterprise in Safaga and Suez-Egypt, huge sea mount and canyon have been discovered in the Red Sea.

On the one hand, Shukri and Higazy (1944) referred to the Red Sea as probably unique to its very irregular bottom topography (excluding the Gulf of Suez), as nature and distribution of its bottom deposits are unmatched in other seas. Jokela (1963) revealed that the Red Sea bottom topography is rough except below the shelves. He distinguished narrow shelves in the northern Red Sea; wide shelves and narrow median valley in the south. Braithwaite (1987) stated that depths of hundreds of meters are generally reached within a few kilometers of the shoreline and there is commonly no effective continental shelf. In the northern Red Sea, Tewfik and Ayyad (1982) referred to the presence of northwest trending negative gravity anomalies, which are believed to represent subbasins. These subbasins are generally parallel to the coastline and separated by NW-SE fault trend that is the main structural element in the area. Other cross fault trend is perpendicular to the former trend. However, Said (1990) assumed that the Red Sea marginal homocline dips gently eastward from the shore and is made up of many horsts and fault blocks. The outcropping sediments along the Red Sea coast have a regional eastward dip to the sea. Dip angles range from 35° at the inland basement contact to nearly horizontal near the shore (Said, 1990). Further offshore, investigation of the seismic data indicates a relatively narrow NW trending zone of pre-middle Miocene horsts and tilted fault blocks. The structural axes are generally shifted laterally, most probably owing to the E-W cross faults. Uchupi and Ross (1986) used seismic reflection profiles to distinguish oblong low relief features (gently folded structures) in the northern Red Sea, which appear to have resulted because of the intrusion of the Miocene evaporites. They divided the rift of the northern Red Sea topographically into narrow shelves, which are less than -50 m bsl, marginal slopes with reliefs of about -600 m bsl and the main trough below -600 m depth. The central part of the main trough consists of northwest trending highs and lows truncated at the north by the Gulf of Aqaba trend. Moreover, submarine terraces and canyons have been distinguished in several localities along the Red Sea Coast, e.g. Sharm El Sheikh, Gulf of Elat, Sudan, and Saudi Arabia (Emery, 1963; Gvirtzman et al., 1977; Fricke, 1983; Reaches et al., 1987). Embabi (2004) inferred from bathymetric lines on topographic maps, scale 1:1000.000 small deltas along the Egyptian Red Sea Coast below the present-day sea level in front of the main wadis. Moawad (2008, 2013) distinguished probable submarine terraces (-100, -50, and -20 m bsl), submerged like-deltas and deep incised canyons of ~500 m depth.

2. Study Area

The Egyptian Red Sea Coast extends southwards from the southern limits of Gulf of Suez (27°43`N) to the Egyptian-Sudanese border (22°00`N), some 790 km (Figure 1). It seems to be straight for the most parts due to the tectonic genesis of the Red Sea itself as a major part of the African Rift Valley. The transition zone between the Gulf of Suez and the northern Red Sea is avoided in this study, not to be confused with the Red Sea topography. The study area is a good example of a passive continental margin as no collision or subduction is taking place and tectonic activity is minimal. It is generally characterized by narrow continental shelves as there is a little sediment transport by the present-day drainage system or redistribute by the longshore currents.

The coast is bounded from the west by the Red Sea Mountains that are mainly formed from igneous and metamorphosed rocks with narrow sedimentary cover.

The mountains overlook the Red Sea by very steep slopes owing to tectonic effects, leaving a narrow coastal plain along most of the Egyptian Red Sea Coast. Erosional, submerged, and emerged features are the most common phenomena. Nevertheless, depositional landforms are patchy distributed and are being formed as a result of erosion and transportation of unconsolidated materials (Moawad, 2008). Presence of ancient coral reefs and submerged wadi outlets (locally known as sharms) indicates the lowering of sea level in the past.

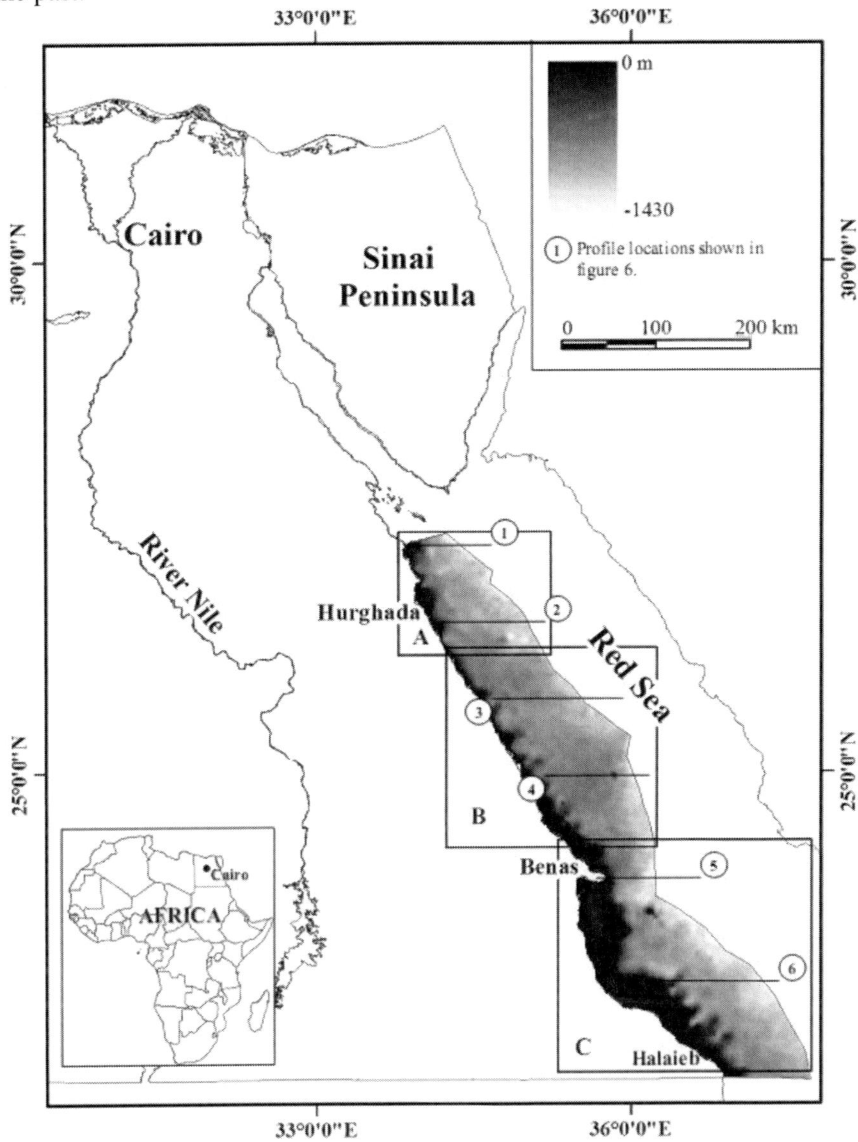

Figure 1. Location and bathymetry of the study area. (Boxed areas showing locations of figure 3).

General oceanography reveals that capillary waves (≤0.6 m) are the most common throughout the year and the averaged wave height ranges from 0.6 to 1.5 m (Moawad, 2008). The wave occasionally exceeds 3 m whereas the wind speed increases above the normal average (6.15 and 4.5 m s^{-1} in Hurghada and El Quseir, respectively, Climatological Normals of Egypt, 1975). It rarely exceeds 5 m during gale wind as the wind speed is +17.49 m s^{-1}, which mostly occurs during winter and early spring (Edwards and Head, 1987).

In addition, the study area spans from the micro-tidal environments. Tide is semi-diurnal, has two high and two low waters per tidal day (Edwards and Head, 1987). Neap tide ranges from 0.83 to 0.6 m in Hurghada and Halaieb correspondingly (Figure 1). Tide amplitude is significantly low; varies between 0.46 and 0.48 m in winter and summer, respectively and the maximum amplitude is about 0.5 m owing to the seasonal changes during the winter months (Behairy et al., 1992). Nevertheless, tidal water inundates the coastal low lands up to a few hundred meters (Moawad, 2008). As the longshore currents are mainly governed by the prevailing winds, they are generally very weak. The average velocity of the longshore currents is about 0.12 m s^{-1} (Edwards and Head, 1987). In other words, role of the coastal processes is mostly insignificant.

The main objectives of this study are to recognize the major submerged and/or submarine landforms that marked the Egyptian Red Sea Coast and the continental shelf. The study may be helpful for further researches, as well to construct a general overview about the geomorphologic characteristics and evolution of the continental shelf and the Juxtaposition landmass.

3. Data and Methods

The study depends on five data sets of topography and bathymetry, which have been integrated and/or overlaid for the purposes of data analysis. The data sets are the global bathymetry data set from Smith and Sandwell (1997), the SRTM DTED® 2, the Soviet Military topographic maps of scale 1:200.000, the US Army topographic maps of scale 1:250.000 and the ETM+ Landsat images.

The global bathymetry data set represents an estimation of seafloor topography. The data were obtained from shipboard depth soundings combined with gravity data derived from satellite altimetry. The final production is a gridded representation of seafloor topography for all ice-free ocean areas within ~72° latitude. The depth data were obtained by screening

6905 surveys from the NGDC (Marine Trackline Geophysics CD-ROM version 3.2), the Scripps Institution of Oceanography and Lamont-Doherty Earth Observatory databanks, and other data using quality control procedures based on those of Smith (Amante and Eakins, 2009). The data were obtained from the interactive database management system GEODAS (GEOphysical DAta System) developed by the National Geophysical Data Center (NGDC) (*http://www.ngdc.noaa.gov/mgg/image/2minrelief.html*). Data are available in ASCII raster grid format including arc header using spatial resolution 2 minutes along latitudes and longitudes (about 4 km at the equator and 2.6 km at 45° latitude).

SRTM DTED® 2 is defined as a uniform grid of elevation values, spaced at three-arc second (approximately 90 m) intervals. Elevations are rounded to the nearest integer meter and are referenced to mean sea level defined by the WGS84 EGM96 (Earth Gravitational Model 1996) geoid. SRTM elevations represent the elevations of the reflective surface (e.g. tree canopy, building roof, bare ground) for the radar return beam and have not been reduced to bare Earth. Accuracy of the SRTM data is 16 m absolute vertical error (90 % linear error, with respect to the reflective surface), 20 m absolute horizontal error (90 % circular error) and 10 m relative vertical error (90 % linear error) (NIMA, 2000). SRTM DTED® 2 was preferred for many reasons i.e. i) the ability of identification, delineation, and elevation determination of water bodies that meet the minimum size criteria, ii) coastlines and sea level are set to 0 m, iii) double-line drains (rivers) greater than 183 m in width can be delineated, iv) islands are delineated if greater than 300 m in length or, for smaller islands (down to14.400 m^2); if at least 10 % of the island's elevations are 15 m or more above the surrounding water. Odd points (spikes or wells) in the elevation data were removed and voided out if they exceeded the mean elevation of the surrounding (eight) neighboring elevation posts by 100 m or more.

Bathymetry lines were firstly digitized from thirteen topographic map sheets produced by the Soviet military of scale 1:200.000 and another eight topographic map sheets produced by the US army of scale 1:250.000. Generally, contour interval varies from 10 m onshore and 100 m offshore. The lines were converted into points using ArcGIS v.10. Next, a raster surface was processed using ordinary kriging interpolation procedure. This procedure is preferred to determine error(s) of the global bathymetry data as the mean altitude varies greatly between the landmass and the surrounding continental margin. Therefore, the procedure attempts to have a mean residual error equal to or near zero and predict the semivariances multiplied by the ordinary kriging weights based on the variance of the vector data (Lloyd, 2007). This was carried out using 4518 control points derived from the topographic maps. Then, these points were input into the geostatistical analyst to create a spherical semivariogram. After that, a prediction surface was created and differences between the global bathymetry data and the DEM derived from the topographic maps were determined along some shared localities. The prediction surface from kriging was subtracted from the global bathymetry data to get a new DEM with the error removed. The mean of the prediction error was estimated at -0.00453 and the RMSE (root-mean-square error) was estimated at 23.28. It is known that these values decrease as the number of nearest neighbors increases (Lloyd, 2007). Finally, bathymetry data (derived from topographic maps and the error removed DEM) were integrated together to create the final digital elevation model using spatial resolution 100 m. Golden Software Surfer 9.11 was preferred for better representation of the data in 3D perspective views.

However, major wadis were extracted from the SRTM DTED® 2 to clarify their relation to other submerged landforms (e.g. submerged deltas and submarine canyons). This was done by determining the directions that water will flow out of each cell to its steepest down-slope neighbor (flow direction) relying on the 8D method (O'Callaghan and Mark, 1984). Flow accumulation function was applied to tabulate for each cell the number of cells that will flow to it (Chang, 2006) and cells of high flow accumulation were used to identify the main channels. ETM+ Landsat images were used to define accurately the pour points (outlets) of the main hydrographs and to visually verify the output streams. Stream network was derived from the flow accumulation layer by threshold value of $\geq$150 cells, which was preferred as the resulting streams are likely corresponding to a network obtained from traditional methods using topographic maps of scale 1:50.000 (Tarboton et al., 1991; Moawad, 2008). Streams were extracted by assigning the flow accumulation layer to integer values ≥ 1 for a stream and no-data for background. Thereafter, streams were ordered according to Strahler (1957). For the purposes of data representation, major and trunk streams are only represented while the lower orders (i.e. orders 1st, 2nd and 3rd) were excluded.

4. Results of Data Interpretation

Major submarine topography in the study area can be summarized as follows (Figures 2, 3):

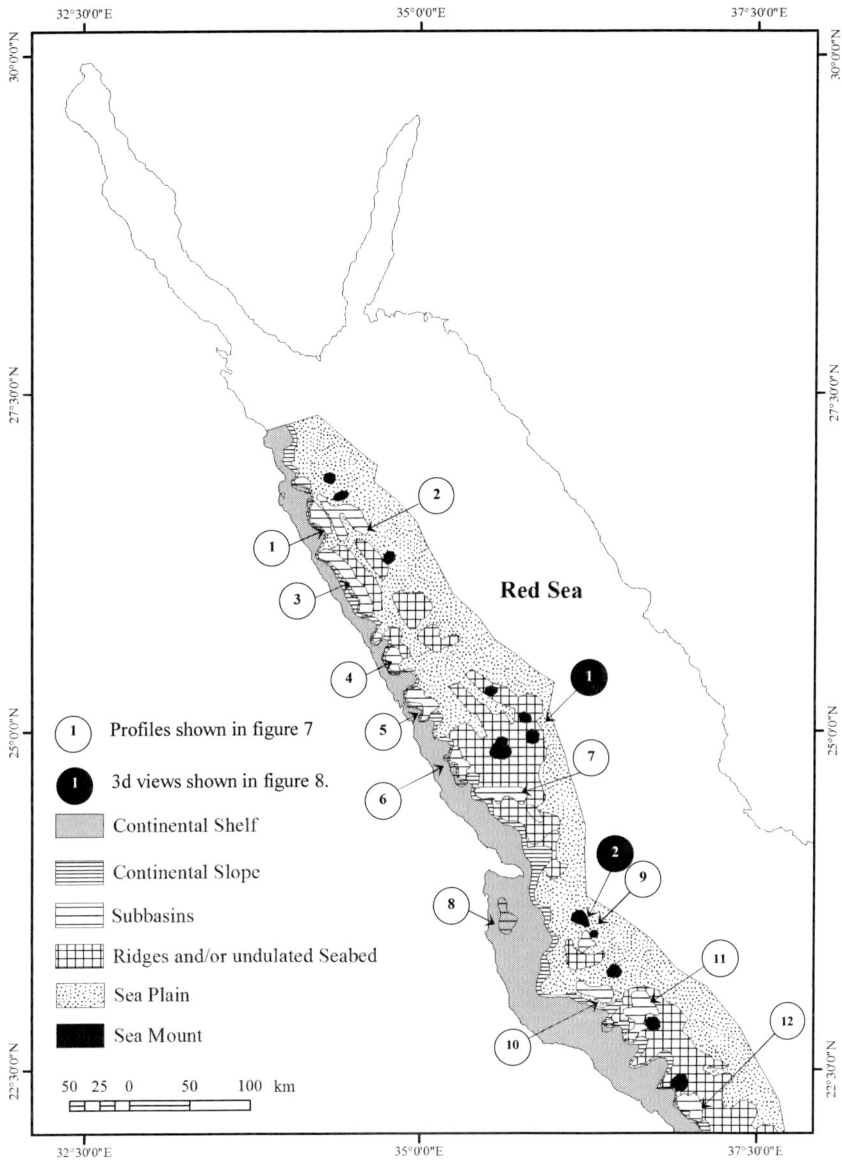

Source: Extracted by visual and digital interpretation of the bathymetric data set.

Figure 2. Major submerged topography along the Egyptian Red Sea Coast.

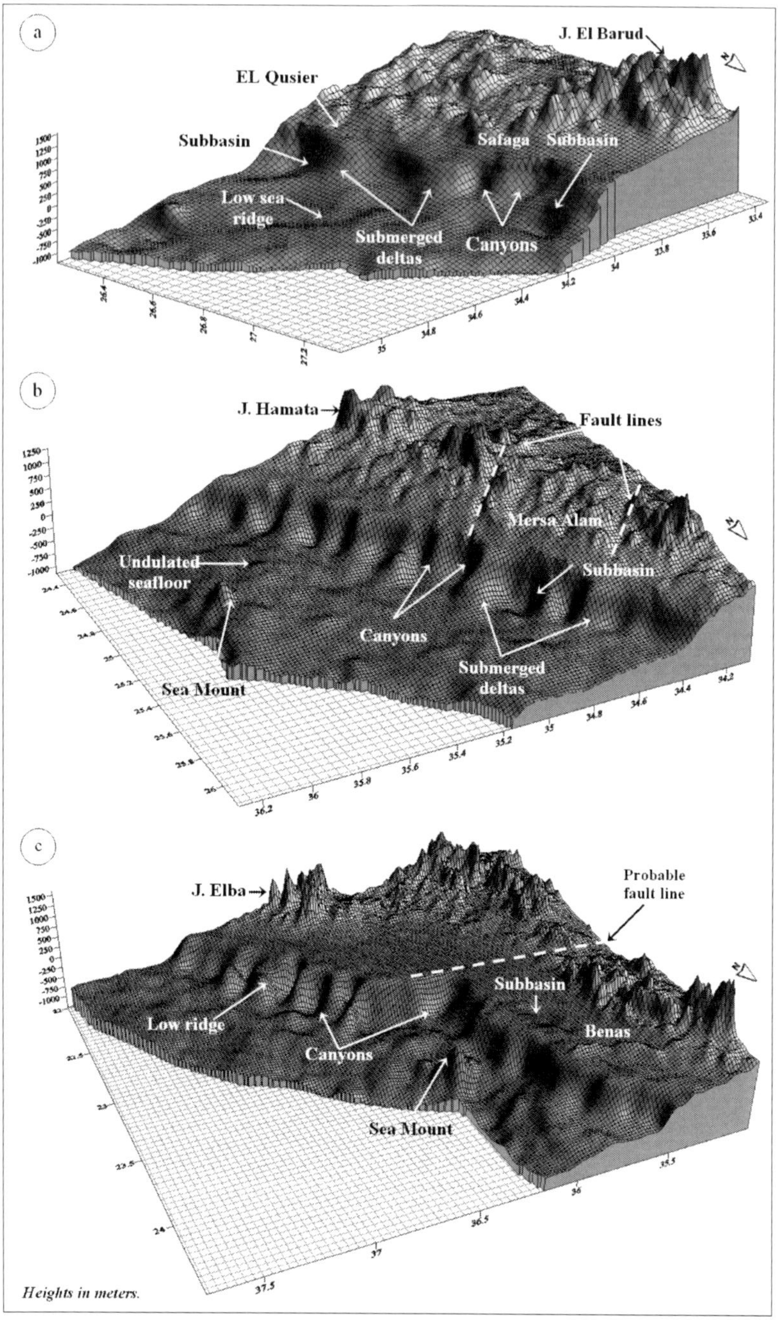

Figure 3. 3D perspective views showing the main submarine topography along the Egyptian Red Sea Coast. (Locations shown in figure 1).

4.1. Coral Reef

The modern coral reefs have developed mainly over the past 5.000 years in response to rising sea level. Red Sea is a favour environment for coral growth. Surface water temperature in northern Red Sea differs from 20° C to 27° C in summer with a general average of 24.9° C (Radwan 2001). Jones et al. (1987) stated that the surface water temperature of the Red Sea may reach over 45° C in summer.

However, salinity of the Red Sea ranges from 39-41‰ in the upper layers (Jones et al., 1987) and about 40-41‰ in Hurghada (Neuman and McGill 1962; Radwan, 2001).

Nevertheless, torrential rains and seasonal flooding by the wadi systems can cause sudden decrease in salinity and severe damage to corals at the wadi mouths. However, corals of the Red Sea are generally embedded under the following coral types:

- *Fringing reefs:* These are the most dominant coral types in the Red Sea that lie very close and parallel to the shore line. They are separated from the shore by narrow, shallow lagoons owing to lack of upward and outward growth (Figure 4). They reveal a platform-like form owing to a combination of constructive and destructive processes. The living coral forms a thin veneer over a platform composed mostly of the remains of previous coral generations. The platform varies generally from few tens to hundreds of metres width and its outer-seaward margin acts as wave breaker, leading to decrease and limiting wave energy. The inter-lagoons are mostly oval or circular in shape of few meters in diameter and depth. They are recognized along the coastline mostly without exceptions.
- *Barrier reefs:* Head (1987) argued that true reefs as defined by Darwin as "steady growing reef systems in subsidence environment ", do not occur in the Red Sea. According to that, barrier-like structures extending over few kilometres. Their distribution depends on the Red Sea tectonism where they are mainly founded on the seaward edges of large fault blocks or on the top of elongated salt diapirs (Head, 1987). Barrier reefs in the Red Sea are known as ridge reefs by Guilcher (1988) as they form elongated narrow framework running in the general direction of the Red Sea.

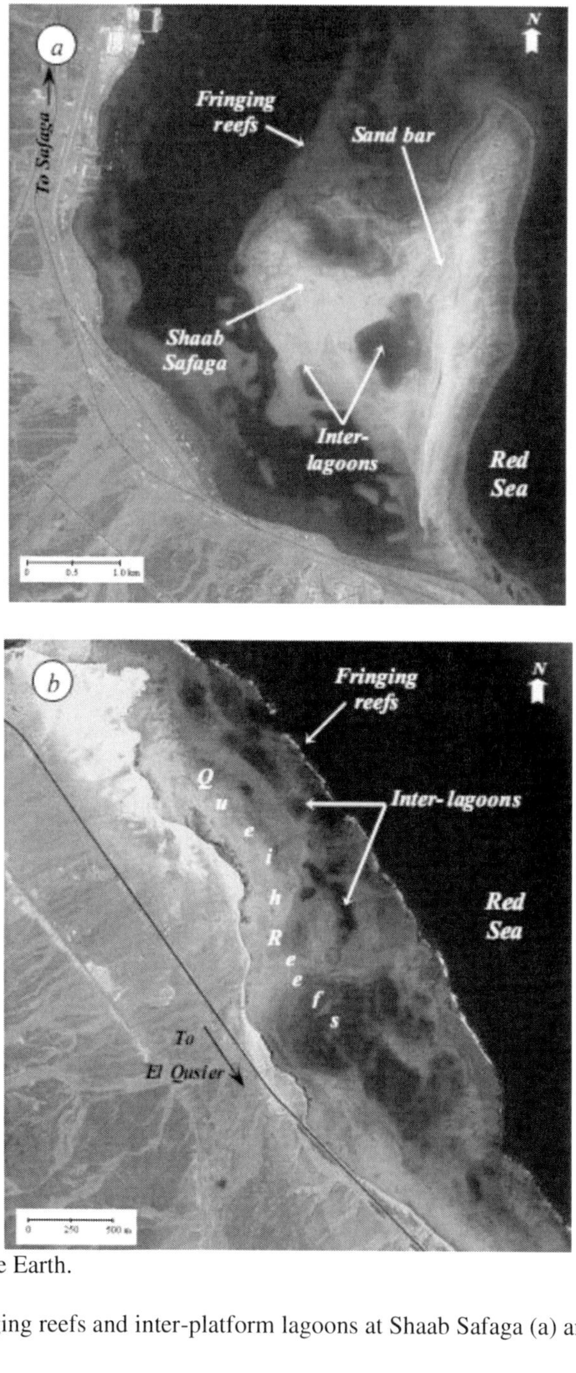

Source: Google Earth.

Figure 4. Fringing reefs and inter-platform lagoons at Shaab Safaga (a) and Queih reefs (b).

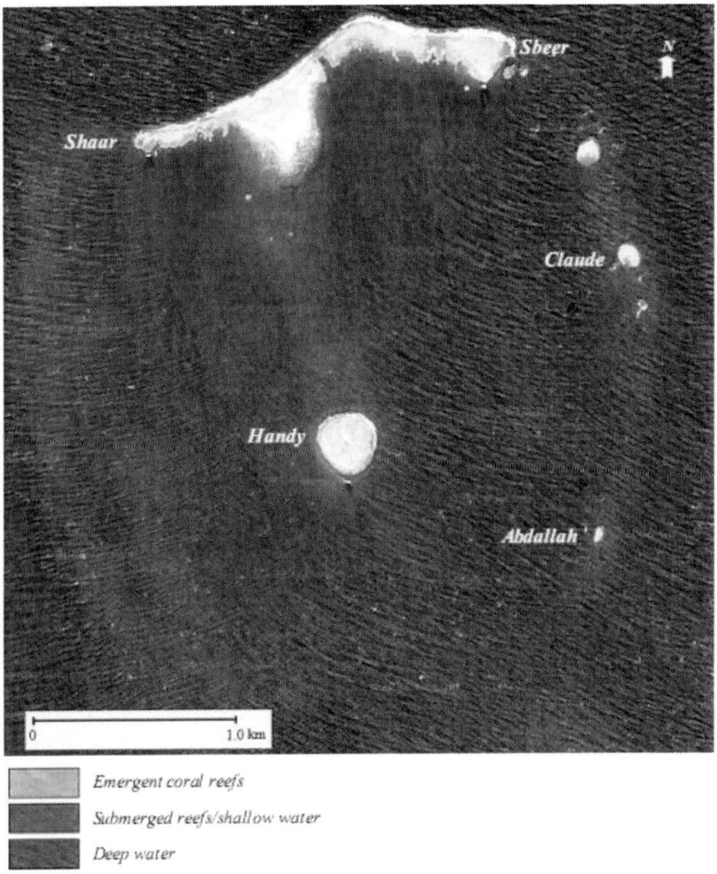

Source: Google Earth, 2007.

Figure 5. Atoll reefs eastern of Shaab Safaga.

- *Atolls* are annular reefs that develop at or near the surface of the sea when islands that are surrounded by reefs subside. Vandersträte (2006-2007) revealed that the formation of such atoll-like structures is associated with smaller salt diapirs and are not well developed in the Red Sea. Nevertheless, a group of reefs (Shaar, Shab Sbeer, Shaab Claude, Handy and Abdallah reefs) reveal atoll-like structure in the vicinity of Shab Safaga. This group of reefs forms actually an annular atoll. Most of its parts are submerged under the present sea level, while Shaar, Shab Sbeer, Shaab Claude, Handy and Abdallah reefs are the most emergent ones (Figure 5). Shaar and Shab Sbeer form the northern emergent rim of the atoll extending west-east generally for about 2 km and their width varies greatly from place to place between 80 to 400 m. Claude and Abdallah reefs are so tiny emergent reefs on the eastern rim of the atoll. On the southern margins of Abdallah reef an Egyptian ferry called Salem Express was founded on 16th December 1991 and this is a proof that the reefs are extending away under sea level. Handy reef lies approximately in the center of the atoll manifesting semi-circular reef. It is mostly connected with Shaar and Shab Sbeer through a submerged bridge-like reef platform. The perimeter of the atoll as a whole is about 11 km covering approximately 9.5 km^2. The longest diameter is about 3.4 km from west to east and the shortest one is about 2.9 km from north to south.

- *Patch reefs:* many reefs without a definite structure are defined as patch reefs. They vary from very small-simple coral to large patches. They reach the low tide level and are formed by aggregation of a few enormous coral colonies.
- *Coral Carpets:* Riegl and Piller (1997) used the term "coral carpet" to distinguish the more or less continuous veneers of coral framework following the existing seafloor morphology in Hurghada.

4.2. Continental Shelf and Continental Slope

The continental margin is the submerged outer edge of the continent. It is generally divided into two subdivisions: the continental shelf and the continental slope. The continental shelf starts from marks of the high water on the shore (littoral zone) and exhibit noticeable breaks in declivity from gentler slope near the shore to one that is much steeper (continental slope) beyond the shelf break. Along the Egyptian Red Sea Coast, depths of the shelf range from a very few meters (-1 m) down to ~120 m bsl. It is mostly covered by coral communities. Width of the shelf changes locally, varying from a few kilometers (5 km) to tens of kilometers (46 km). The widest part ranges from 24 to 46 km in the area that extends from Ras Benas to the southern tip of El Foul Bay (15 km long) (Figure 6).

The continental shelf can be considered as tectonic and to a lesser degree depositional margin. Its accretion is constrained primarily by sediment supply since role of the coastal processes is mostly insignificant (Moawad, 2008). However, deposition is predominantly restricted to submerged deltas and some reef communities.

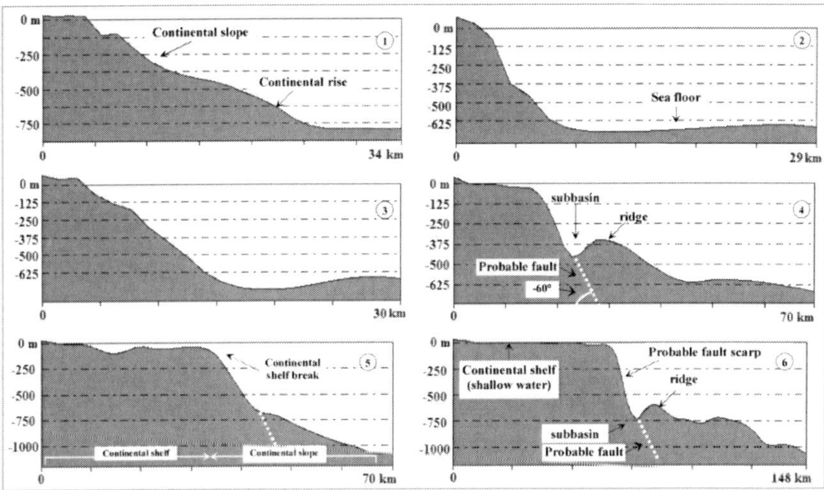

Figure 6. Topographic cross sections showing the main physiography of the continental shelf. (Locations shown in figure 1).

Topography of the continental shelf is not completely flat. The present-day platform slopes eastward gradually (2°–9°) between the mean low water to the further limit of the breaker zone toward the offshore. It is mostly covered by coral reefs that run parallel to the coastline (Unlimited Media, 2004). The breaker zone is defined by the outer limit of the fringing reefs as it acts as a line of breaker. Therefore, most of the waves are forced to be broken in the vicinity of the coastline between 50 and 400 m in average at depths 0.5–1.5 m bsl. The platform is often disturbed by inter-lagoons (pool-like) or mud flats as previously mentioned.

Another interesting feature of the shelf is St. John's cave at Mersa Alam that is reported by many divers in St. John's reefs. The cave is a slightly misleading name it seems since scientifically it seems to be one of the many fissures that marked the continental shelf at depths of -6 to -25 m bsl. Such incised fissures are known by leader divers as tunnels or passages, of which many fissures have no direct access to the sea surface, although there are always small openings to allow light penetration. The sea level dropped approximately 100 m below its present-day level during the last glacial maximum (LGM) (~23– 18 ka bp, Gasse, 2000). It is assumed that a broad part of the shelf was exposed and subjected to subaerial erosion and resulted in the development of Karst forms beside shallow canyons and the present-day embayments (locally known as sharms) as the wadi system carved the continental shelf. This may explain the presence of the present-day coastal lagoons and some other small-scale sinkholes and submerged caverns. Nowadays, development of the modern reef communities led to shallow the shelf with numerous shoal reef areas that act as an extensive trapping of sediments. Regional tectonic pattern of the continental shelf is mainly influenced by the surface faults extending eastward into the main Red Sea depression. Tewfik and Ayyad (1982) revealed that the Red Sea marginal homocline dips gently eastward from the shore and it is made up of many horsts and fault blocks. Further offshore, seismic reflection data indicate a relatively narrow NW trending zone of pre-middle Miocene horsts and tilted fault blocks. Ahmed (1972) revealed that in late Pliocene and Pleistocene times there was renewed block faulting accompanied by igneous intrusions. As a result, the Red Sea depression is characterized by rectilinear faults bordering blocks, which were rising or sinking concurrent with the Neogene deposition. In many cases the structural axes are shifted laterally, probably as a result of E–W cross faults (Tewfik and Ayyad, 1982). The continental slope starts from the edge of the continental shelf (shelf break) and declines progressively down to 500 and/or 700 m bsl. The slope is generally narrow (~1 km width) where it descends abruptly toward the deep seafloor and looks mostly like a fault-scarp. The form of the continental slope exposes normal faults trending NW–SE matching the main trend of the Red Sea trough with a fault plan of ~60° anticlockwise and steep slope (Figure 6). Said (1990) cited that the major faults are attributed to the initial rifting of the Red Sea and they experienced tectonic events at the edge of the continental shelf (e.g. uplift, volcanic activity) in the later stage. Sometimes width of the continental slope attains about 8 km as a result of the submarine deltaic deposits descending into the seafloor. No distinctive gently-sloping transition (continental rise) between the continental slope and the deep seafloor was distinguished. This may be due to lack of thick accumulation of sediments from the neighboring landmass that is transported to the shelf and then down the continental slope. Generally, the continental slope has been interrupted by folding, faulting, and erosion with the present shelf configuration resulting from Pleistocene subaerial erosion with very little Holocene deposition (Said, 1990).

4.3. Subbasins

Based on gravity data Tewfik and Ayyad (1982) recognized many subbasins parallel to the coastline and separated by structural ridges. Using digital and visual image interpretations 12 subbasins were recognized from the digital elevation model. They vary in shape between longitudinal and semi-circular (Figure 2).

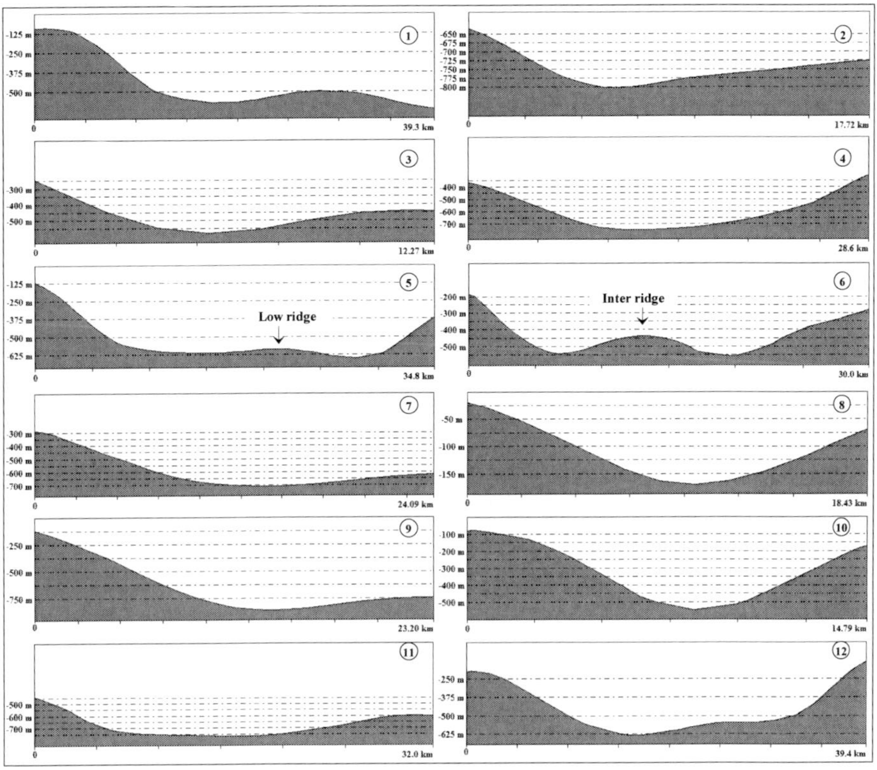

Figure 7. Topographic cross sections of the subbasins. (Locations shown in figure 2).

They run parallel to the continental shelf except the subbasin numbers seven and 11 that run from east to west. Generally, the subbasins are restricted to the lower tip of the continental slope from the east and to the sea ridges from the west with the exception of the subbasin number eight that lies on the continental shelf near El Foul Bay. Depth of the subbasins ranges generally between -196 m (subbasin number eight) and -853 (subbasin number nine). They vary as well in width from 12 to ~40 km and the general physiographic settings reveal that the western limbs are generally higher than the eastern and southern limbs (Figure 7). Some subbasins are separated by low ridges, which seem to be a part of regional undulated seabed strata of a wide extent. Tewfik and Ayyad (1982) assumed that the subbasins seem to have been formed during the development of the regional arching and subsequent collapse of the

Red Sea. They suggested that they are probably filled with middle Miocene evaporites and younger continental sediments.

4.4. Sea Ridges

Ridges are common in the vicinity of the lower tip of the continental slope mostly without exceptions. Two types of ridges can be distinguished based on their altitudes. The first one is the low ridges that rise a few tens of meters above the surrounding seafloor (Figures 3 and 6). They range in elevation from -400 to -600 m bsl. The second is the higher ridges that exceed +100 m above the surrounding seafloor (Figure 6). Summits of the ridges are generally round and differ greatly in depth from -300 to -600m bsl. The lower base of the ridges differs as well locally, ranging from -500 to -700 m bsl. Based on their symmetry, they might be inferred as a result of a regional fold system. In the vicinity of the continental slope, probable normal faults separate the sea ridges from the continental slope with an estimated dip of a fault plane of -60°. Investigation of the geological maps revealed that some offshore islands (e.g. Safaga, Sernaka Island, Mukawwa, Zabargad and the Rocky Island) are built up of the same rock formation(s) as the neighboring landmass (sedimentary, igneous and/or metamorphosed rocks). The islands constitute the small merged areas that are standing from the top of the sea ridges owing to the regional upward of the continental margin. Marginal upward might be a result of fault reactivations and major folding during the formation of the Red Sea trough. The islands are therefore tectonic and not volcanic in origin. Jokela (1963) suggested that episodic compression occurred in the direction of the Red Sea axis in pre-Oligocene time and was expressed by folding of soft sediments along the Egyptian shelf and possibly by uplift of mountains in Turkey. Simultaneously, tension perpendicular to the sea axis formed fractures parallel to the axis, some of which served as volcanic and hydrothermal vents. Marshak et al. (1992) revealed that Zabargad Island was affected by the final phase of tectonism resulting in an array of high angle faults and associated folds. Together resemble palm structure typical of cross fracture zones in rift. The southwest corner of the island was subjected to large recumbent folds. It appears to be an uplifted fragment of sub Red Sea lithosphere and it is essentially the result of upward motion of mantle derived ultramafic bodies (Bonatti et al., 1983).

4.5. Seamounts and Knolls

Seamounts are other distinctive features within the large structurally complex segment of the Red Sea. Twelve seamounts were recognized, which rise surprisingly from the seafloor and peaking below sea level (not reaching to the water's surface). They range in elevation between -100 to -800 m bsl and have mostly circular shapes, which seem to be a result of small volcanic cones. Seamounts number s one and twoshown in figure 2 in specific are distinguished as independent features since they exhibit typically conical topography with very steep sides. They rise abruptly from the sea floor of -600 and -800 m depth, respectively up to -350 and -450 bsl in the same order, and seem to be extinct volcanoes with round peaks (Figure 8).

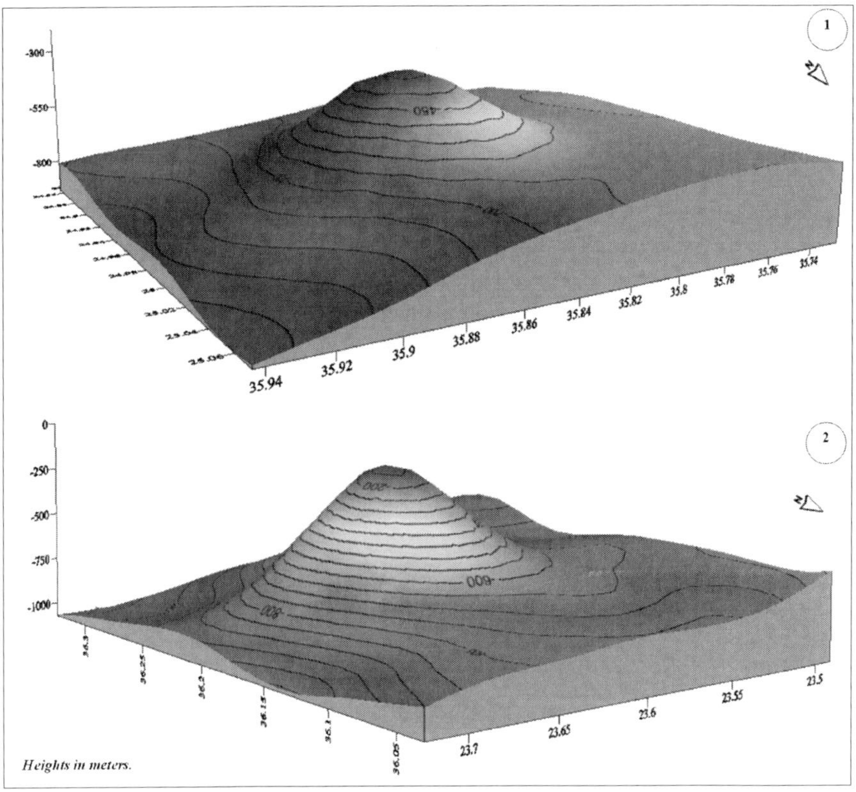

Figure 8. 3D perspective views showing the sea mounts. (Locations shown in figure 2).

Other minor features could be distinguished as sea knoll (seamounts with flat top) as they rise gradually from the undulated seafloor, mostly due to the regional folding system and upthrust faulting. These knolls play an important role in exposing tips of some reef pillars as semi-atoll or off shore islands such as Al Akhawein islands, which are situated 67 km offshore east of El Quseir. The two islands are likely the exposed tips of two massive reef pillars that rise from the sea.

The pillars cover the pinnacles of two undersea knolls rising from about -300 m bsl (Moawad, 2008). More recently, a huge sea mount has been discovered east of Suez, Egypt by the British survey ship "HMS Echo" using hi-tech sonar. The mount rises to 40 m below the surface of the Red Sea (Figure 9) (Maritime Reporter and Marine News Magazines Online, 2004).

Figure 9. 3D perspective view shows sea mount discovered by HMS Echo in the Red Sea (East of Suez, Egypt). (©The U.K. Royal Navy).

4.6. Submarine Canyons

The submarine canyons are the most conspicuous erosional forms marking the outer continental margin along the Egyptian Red Sea Coast. They carve the deep fissures along the continental slope, which are probably produced by an ancestor drainage system. Shepard and Dill (1966) argued that such canyons appear to have had several origins and producing distinctive types. They can be distinguished within the study area on the basis of their morphology as follows:

4.6.1. Deep Incised Messinian Canyons (Grand Canyon-Style)!

Deep incised canyons are the most common features that significantly break the continental slope. They reach ~500 m below the present-day sea level and distinguished by remarkable steep walls that seem to have been cut in the bedrock (Figure 10). In some cases, head of the canyons may be traced to the upper part of the continental slope at depths -100 and -150 m bsl. Longitudinal profiles of the canyons are generally concave and sloping strongly seaward from the head of the canyons to the deep sea. Although, the incised canyons sometimes follow the major structural features seaward from the adjacent land as shown in figure 3, they seem to be the result of a great geological event that influenced the continental margin since they are common along the Egyptian Red Sea Coast mostly without exceptions. Moawad (2008) showed some incised canyons spreading along the coastal area extending between Safaga and El Quseir. In February 2013, the U,K, ministry of defense announced that a canyon more than 243 m deep has been discovered on the floor of the Red Sea about 250 miles north of Safaga by the U.K. Royal Navy's HMS Enterprise using an echo sounder. The canyon is known as Grand Canyon-Style ocean floor (Figure 11). The commanding office of the Enterprise stated that the canyon could have been created by ancient rivers cutting through the rock before the area was flooded and became the Red Sea (Thompson, 2013). Siroko (2008, personal communication)* believed that they are a result of the late-Miocene Messinian salinity crises (Messinian Event!!) owing to the sudden disconnection of water supply between the Atlantic Ocean and the Mediterranean sea. The Mediterranean water level dropped down to 1500 m below its present-day level, entailed a considerable increase of salinity and intense erosion of all Mediterranean rivers such as the Rhone and Nile canyons. The erosive phase is supposed to have occurred between 5.59 and 5.50 Ma and followed by deposition of non-marine sediments between 5.50 and 5.33 Ma (Krijgsman et al., 1999). Stoffers and Ross (1974) argued that there was a large relief differences between the Red Sea depression and the surrounding land during Miocene time. Therefore, the Red Sea depression received marginal clastic sediments (2-3 km thick) interfingering with evaporites (3-4 km thick) and the Red Sea depression continued to subside (Ahmed, 1972; Coleman, 1974). In context, Gillmann (1968) inferred rapid erosion at the Yemen-Hail and Ethiopian arches during Miocene.

---

* Prof. Dr. Frank Sirocko: department of climate and sediments "Johannes Gutenberg-Universität-Mainz, Germany (personal communication).

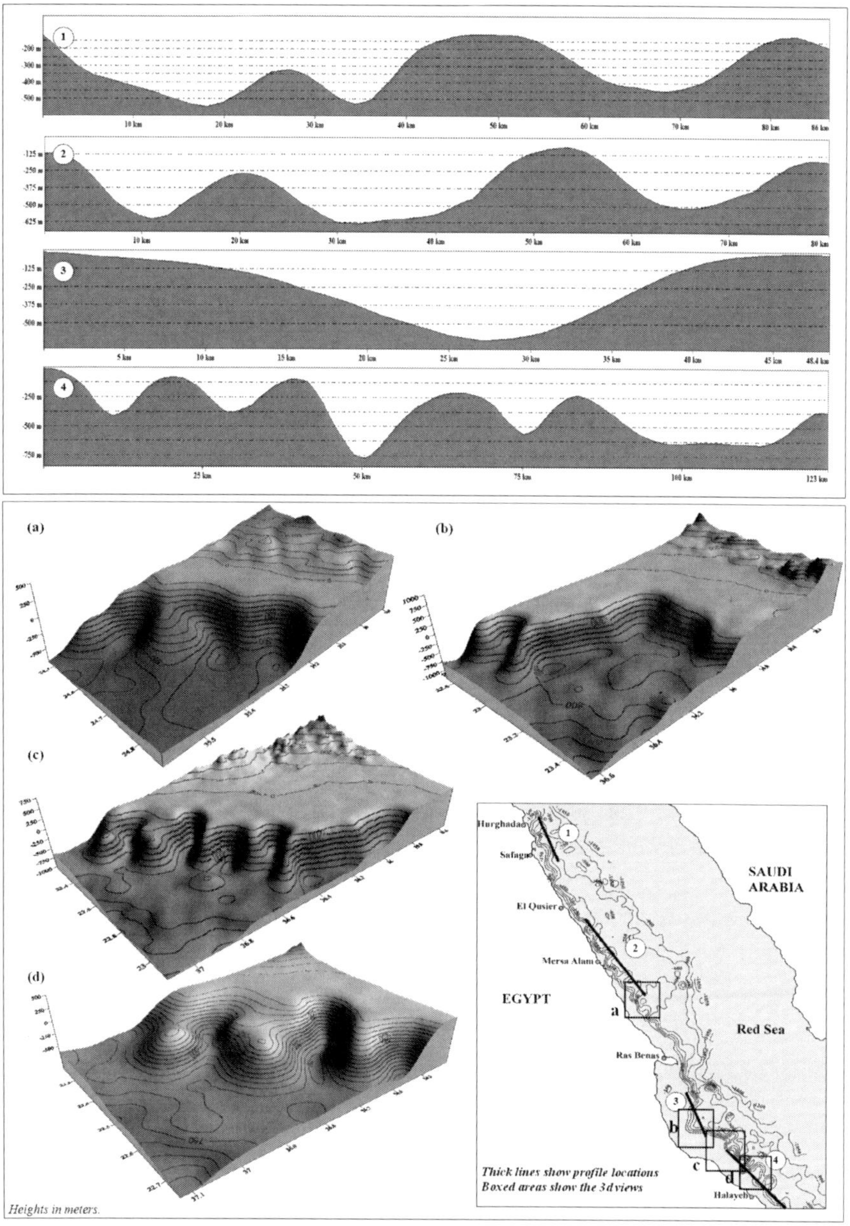

Figure 10. Examples of the deep incised Messinian submarine canyons.

It can be deduced that an old drainage system was the major carrier of such sediments, especially during middle-Miocene time as the climate was warm humid (Khedr, 1984). As a result, a great bulk of the Tertiary sedimentary rocks was swept out from the Miocene highlands in the Eastern Desert of Egypt. Sedimentologic evidences obtained from Jabal El Rusas (25°11'N - 34°46'E) clarified that the lower unit unconformably overlies the basement rocks. The unit consists of 2-3 m orthoconglomerate with angular and round clastic, which range from large boulders to granular. Khedr (1984) assumed that sedimentation of this unit begun in middle Miocene. Small pockets of the Tertiary sedimentary rocks are still preserved in the morphotectonic depressions in the Central Eastern Desert of Egypt such as Jabal Duwi (26°11'N- 34°E) (Moawad, 2008).

Although, role of the recent fluvial process is insignificant, Fricke (1983) denoted that flash floods often of catastrophic nature, which influence the Red Sea region, carry large amounts of sediments into the sea. Accordingly, he suggested that the occasionally gravity flows fed initially by sediments from aligned wadis are a major agent in cutting the deep submarine canyons into steep edges of the narrow Red Sea shelves.

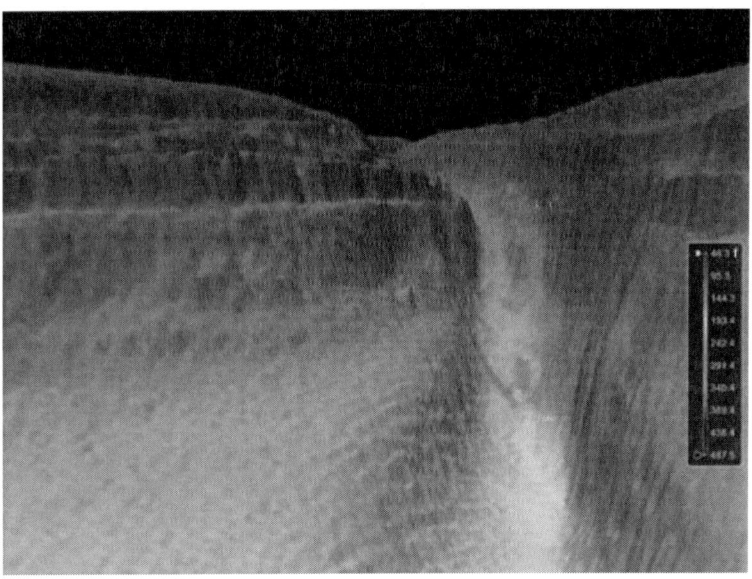

Figure 11. A Grand Canyon-Style on the floor of the Red Sea (South of Suez, Egypt) rendered in 3D and discovered by the UK Royal Navy's HMS Enterprise. (©The U.K. Royal Navy).

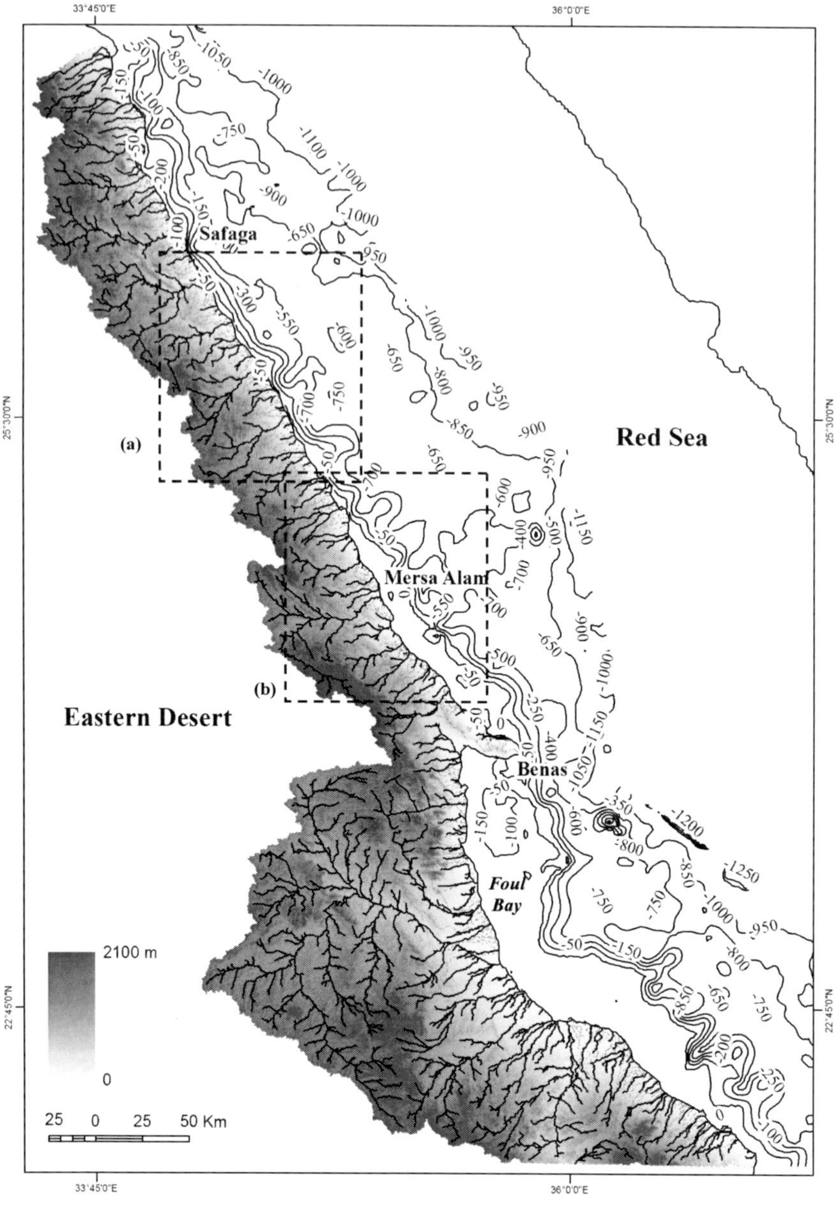

Figure 12. The close relationship between the present-day Red Sea drainage system and the submarine canyons. (Extracted from the DEM, boxed areas approximately show locations of figure 15).

Figure 12 shows a close relationship between the present-day drainage networks and the bathymetry. It is notable, for example that many wadis are debouching into El Foul Bay. The wadis are located some 25 km westward of the head of El Foul Canyon. The major trend of the streams and their tributaries is of W-E orientation. The main wadis range from 100 to +200 km long and their heads are incised in the basement rocks. Another evidence obtained from Jabal Abu Shaar (27°22'N-33°35E) and Jabal El Rusas (25°11'N-34°46'E) in the Eastern Desert of Egypt revealed that the present-day drainage system is typically superimposed (overprinted) on the basement rocks. Therefore, the present-day wadi system is mostly inherited from an older one that may be back to pre-middle Miocene as the climate was shifted in Oligocene toward a Sudano-Guinean type of savannah (Le Houérou, 1997).

4.6.2. Submerged Shallow Canyons

The submerged lower water courses of the wadis exhibit sometimes small-scale canyon-like shapes, which marked mainly the continental shelf and, to a lesser extent, the continental slope. The canyons are relatively shallow as the depth varies from a few meters to tens of meters (-2 to -70 m bsl) and forming estuaries-like (locally known as Sharms). They are characterized by more or less symmetrical winding V-shaped and/or U-shaped (Figures 13, 14). Some smaller-scale tributaries, which have been identified from the hydrological model, are entering the canyons from both sides. The resemblance of submarine canyons to river-cut canyons on land and the juxtaposition of many land and sea canyons has always provided a strong argument that at least the heads of the submarine canyons were originally cut by rivers (Shepard and Dill, 1966). Morphometry of such shallow canyons reveals that they are probably developed during the Pleistocene lower sea level some 90-130 m below the present sea level (Reaches et al., 1987). As a low raised beach borders the present-day coastline (~1.5 m), it can be inferred that an episode(s) of cut and fill related to the sea level changes influenced the shallow canyons. Role of subaerial erosion as well was proposed by Gvirtzman et al. (1977). He assumed that the lowest erosional base level was -120 m bsl in the Red Sea during the last glacial maximum (LGM).

It is worthy mentioning that water circulation of the estuaries (Sharms) is essentially controlled by water salinity. The inshore salinity of the Red Sea may rise dramatically and have given rise to the terms hyperhaline[1] (80-180

[1] Hyperhaline refers to water that contains more than 40 parts per thousand of dissolved salts. Water with salinity greater than normal sea water is called hypersaline.

‰) and hypersaline (80-300 ‰) as a result of little interchange between the open sea and the coastal lagoons or pools (Por, 1972; Jones et al., 1987).

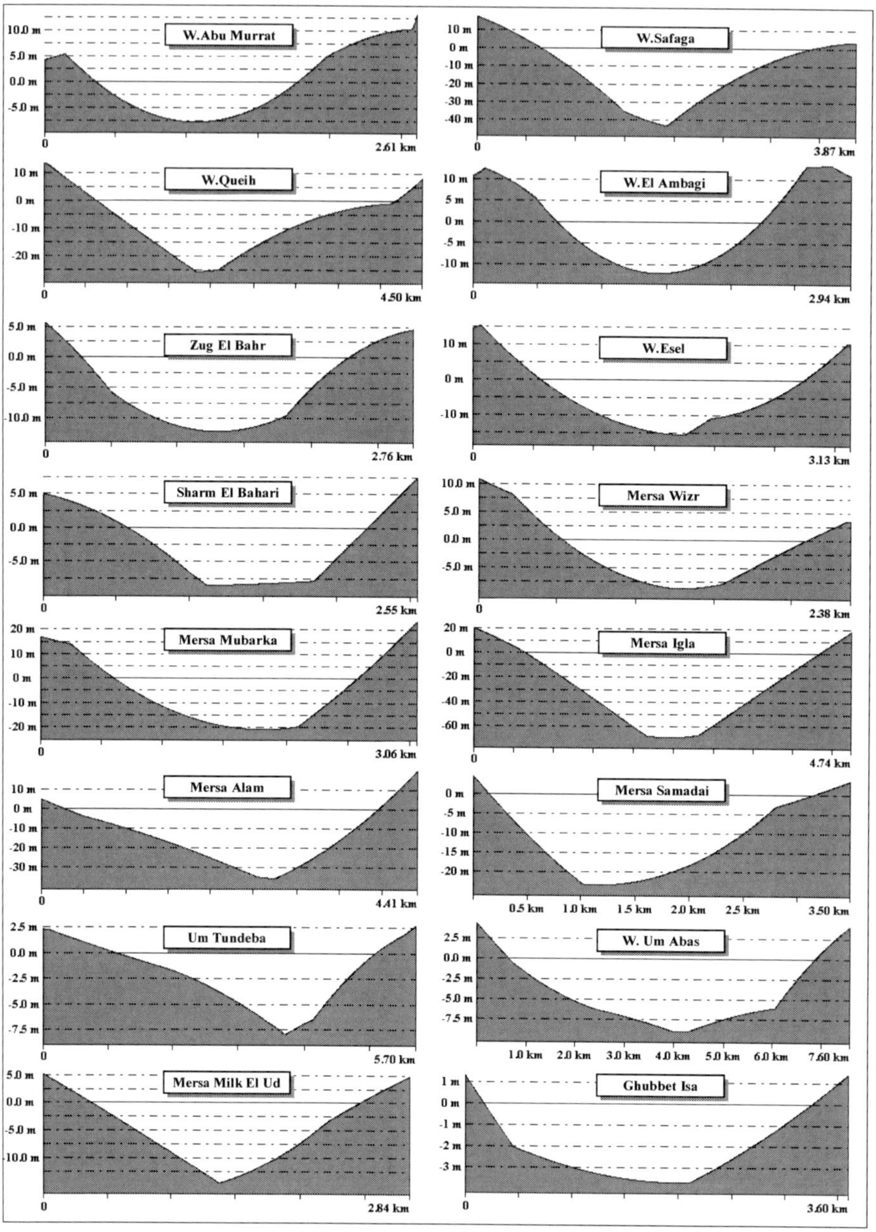

Figure 13. Shapes and depths of the submerged wadi canyons (Sharms).

Figure 14. Sharm wadi Umm Ayish. (Moawad, 2008).

Nasr (1955) demonstrated that the salinity of the coastal pools in Hurghada is both high and variable. According to that, the extreme range of salinity in tide-pools ranges from 41.29 ‰ to 44.09 ‰. At the same time, salinity of the open sea ranges from 39-41 ‰ in the upper layer. Therefore, estuaries of the Red Sea could be classified as inversed estuaries as a result of fresh water scarcity and high evaporation. Surface salinity does not decrease from the open sea to the inner-estuary; instead, water loss from evaporation leads to increase salinity as the inner end of the estuary is approached. This is resulted in increasing density and sinking of high salinity water at the inner end and the water moves inward (landward) at the surface and seaward at the bottom consequently (Tomczak, 1996-2000).

In general, position and configuration of the submarine canyons are controlled by multiple factors, including structural fabric, tectonism, sea-level variations, and sediment supply (Laursen and Normark, 2002). However, idea of marine origin of the canyons as supposed by Shepard (1972) is improbable in the Red Sea as no large active marine erosion has been found throughout the geological evolution of the Red Sea. Even the idea of turbidity current that suggested by Daly (1936) is unlikely since there is no powerful currents in the Red Sea, as far as known to date. It is believed that the role of marine erosion might have been restricted to some form modifications.

4.7. Submerged Deltas

The general configurations of the bathymetric lines reveal submerged fan-like shapes of different sizes either partly or fully extended as a part of the continental shelf. The fan-like shapes are located seaward of the present-day drainage system.

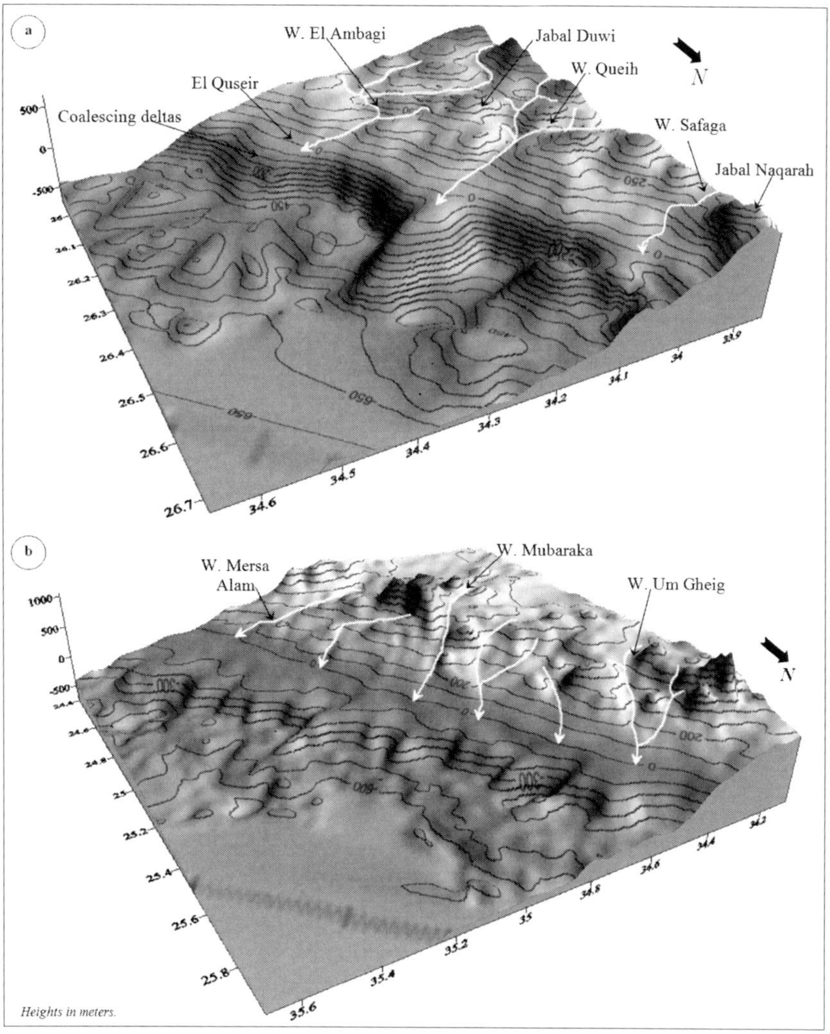

Figure 15. 3D perspective views showing the inferred submerged deltas related to the main present day drainage system. (Locations shown in figure 12).

By following some criteria of the alluvial fan definitions (e.g. morphology of the bathymetric lines, location at topographic breaks), Embabi (2004) and Moawad (2008-2013) assumed that they represent submerged deltas of an ancestor Red Sea drainage system in the Eastern Desert of Egypt. Toes of the deltas are plunging down into the deep sea for ~500 m below the present-day sea level (Figure 15). Individual deltas are bounded laterally by incised canyons, troughs, channels, and therefore the bathymetric lines became more crenulated. The canyons carving the submerged deltas play a significant role in sediment transport mostly during catastrophic flash floods. Sometimes, the lateral boundaries are less distinctive between the adjacent deltas as in the case of delta wadi El Ambagi that seems to be a result of some coalescing deltas.

Correlation between individual wadis is not fully conformable, as the present-day drainage system might have been subjected to meandering, shifting during the Pleistocene lower sea-level and/or tectonic events. However, some prolonged channels have been traced in the shallower part of the shelf such as wadi Queih (Figure 15a) that seems to have been cut the continental shelf during the maximum low stand of sea-level. The landward side of the delta is found just beyond the point where the recent trunk stream of wadi Queih debouching into the Red Sea.

Generally, the submerged deltas seem to be constructed from consolidated alluvial deposits, as processes of diagenesis that resulting in compaction, cementation, and lithification require millions of years to transform the sedimentinto sedimentary rocks (National Research Council, 1996). Stow et al. (1985) and Kolla and Macurda (1988) determined the factors controlling the development of such submarine system into two groups. The first group includes the primary controls i.e. type of continental margin, relief and tectonic setting of the hinterland, continental shelf-slope relief, nature of basinal areas and basin size, distance from the source region, width and gradients of the shelf-slope, and type and amount of sediment supplied. The second group includes the secondary controls, i.e. climatic nature and vegetation of the source area, and high rainfalls during sea-level changes.

Marine terraces are one of the most important recognized features that marked the submerged deltas as flat surfaces of different extents. During Pleistocene and Holocene, the Red Sea experienced dramatic eustatic changes resulting in a sequence of raised coral terraces and alluvial terraces. Moawad (2008) distinguished three submarine terraces at -20, -50, and -100 m bsl, which are mostly a result of the last glacial maximum (LGM) low sea-level (~100-120 m bsl). It was assumed in this study that the terraces are in the agreement with the generalized curve of sea level rise, which has been proposed by Gornitz (2007), since the last ice age. Based on the analysis of the bathymetric data, notable submarine terraces were found in different locations and levels along the coast as shown in table 1 (Figure 16). They constitute mostly flat narrow surfaces and differ locally in depth and width.

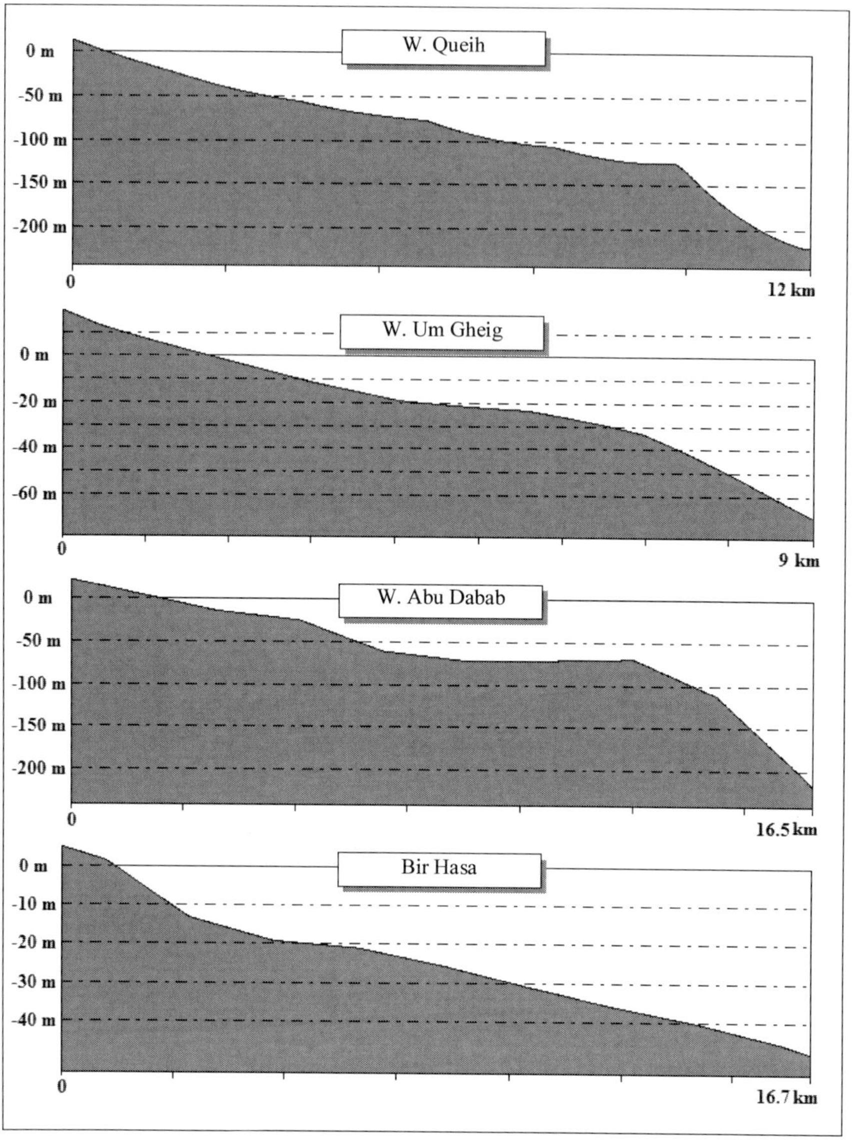

Figure 16. Topographic cross sections showing the main physiography of the submarine terraces below the present-day sea-level.

In comparison with the generalized curve of sea level rise since the LGM (~23-18 ka bp), it can be inferred that the terrace 100-120 m bsl may be in correspondence to the last glacial maximum (LGM) in the age range of melt-water pulse (MWP) 1A0 and 1A (~19-16 ka bp). The sea level rose rapidly about 10-15 m in less than 500 years during the MWP-1A0 and followed by another increase about 16-24 m during the MWP-1A. Terrace 50-75 m bsl is probably in accordance with the melt-water pulse-1B (11.5-11 ka bp) when the sea level may have jumped by 28 m. Terrace 20-25 m bsl is mostly formed during the fourth melt-water pulse (MWP) 1C (8.2-7.6 ka bp) that is in accordance with the Flandrian Transgression (Figure 17). It is believed that the shallow submarine canyons were probably formed during the LGM and the present-day embayments were submerged during the Flandrian Transgression.

Table 1. Locations of some submarine terraces below sea-level

Location	*Coordinates*	*Terraces (m)*
Sharm El Arab	26°58'26"N-33°55'2"E	20
W. Naqarah	26°41'1"N-34°00'9"E	20-25
W. Safaga	26°38'27"N-33°59'40"E	20, 50, 100
W. Queih	26°21'15"N-34°09'57"E	50-75, 100-120
W. Abu Hamra	26°17'7"N-34°12'14"E	20
Um Gheig	25°43'57"N-34°33'26"E	20-25

Table 2. Altitudes of the fluvial terraces along the Egyptian Red Sea Coast

Wadi	*Coordinates*	*Heights (meters above the wadi floor)*
Safaga	26°36'18"N-33°56'43"E	2.5-3, 6-7
Gasus	26°33'45"N-33°59'41"E	5-6; 12-13
Abu Safi	26°26'33"N-34°05'4"E	5-6
Abu Hamra	26°16'19"N-34°10'1"E	6-8; 15
Hamraween	26°14'42"N-34°10'38"E	5-6; 13-15
Queih	26°21'5"N-34°06'44"E	10-12, 15-16
El Quseir El Qadim	26°09'12"N-34°14'36"E	5-6
An-Nakheil	26°08'29"N-34°08'26"E	1-1.5
El Ambagi	26°06'24"N-34°14'25"E	1-1.5; 4.5-5; 10-12; 12-21; 30-40; 60
Um Khariga	25°3'48"N-34°50'0"E	1-2; 6-7; 12-14
Seifan	25°6'5"N-34°51'38"E	2-3; 4; 6; 10
Mubaraka	25°30'32"N-34°38'48"E	1-2; 8-14; 20-30
Um Gheig	25°43'16"N-34°32'24"E	1-2; 5-8; 13; 20-30

Source: Arvidson et al. 1994 and field work.

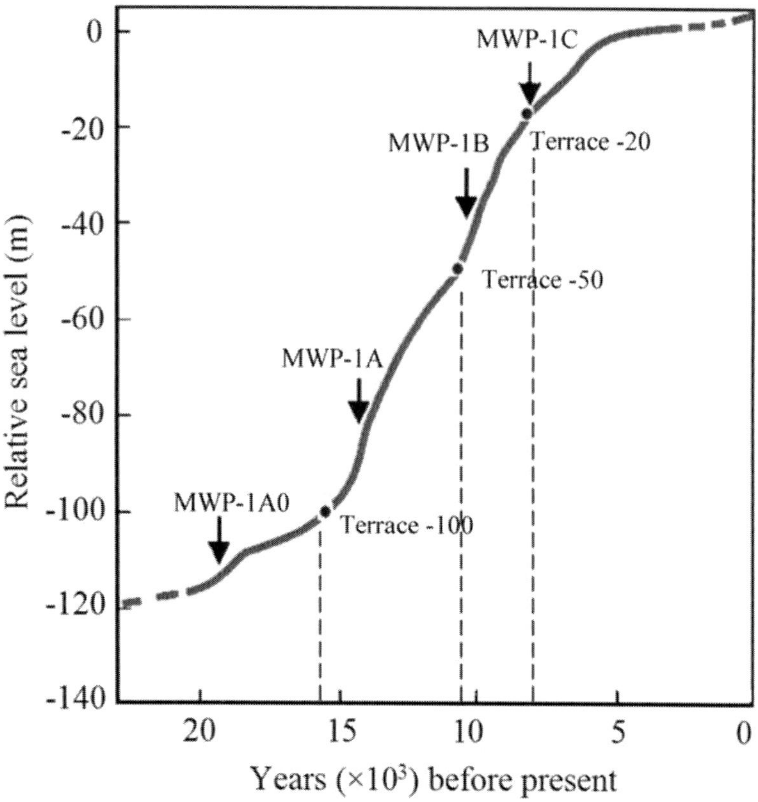

Source: Gornitz, 2007 with minor modification.

Figure 17. Relative correlation between the submarine terraces along the Egyptian Red Sea Coast (100, 50, and 20 m bsl) and the generalized curve of sea level rise since the last ice age.

The submarine terraces in the study area are in accordance with the submarine terraces recorded in southern Sinai at 120, 60, and 20 m bsl (Gvirtzman, 1994). Terraces 100-120 and 50-75 m bsl are in the agreement with terraces 90 and 60 m bsl in Sharm El Sheikh and Ras Burka (Fricke, 1983; Reiss and Hottinger, 1984) and terraces of Elat (90-50 m bsl) (Reaches et al., 1987). Terraces of Elat have likely developed between 70-50 ka bp, when the sea level was 100 to 60 m below the present sae-level (Reaches et al., 1987).

Evidences obtained from the neighboring landmass clarified that the wadis cut through the present-day alluvial fans exhibiting several pairs of alluvial terraces that are in correspondence with the global Pleistocene eustatic sea-level changes (Table 2). It is worth mentioning that the peripheral parts of the NW Red Sea rift were inactive during the last 150-120 ka bp (Hoang and Taviani, 1991; Arvidson et al., 1994; El Moursi et al., 1994; Plaziat et al., 1995; Orszag-Sperber et al., 2001) and the coast attained its present-day shape mostly after the Flandrian Transgression (8-6 ka bp, Goudie, 2004).

Discussion and Conclusions

A long time ago the Red Sea was only known by small-scale bathymetric, magnetic anomaly maps and a few seismic reflection or refraction profiles. Therefore, detection of the major submerged coastal features was unattainable. In this study submerged features along the Egyptian Red Sea Coast from latitude 27°43'N to the Egyptian-Sudanese border at latitude 22°00'N were identified based primarily on the global bathymetry data set from Smith and Sandwell (1997), bathymetry obtained from topographic maps of scale 1:200.000 and the US army topographic maps of scale 1:250.000. The Surface created from the topographic maps has many gaps owing to lack of detailed bathymetric lines. Therefore, the global bathymetry data were integrated to overcome the problem of lack of information. Spatial resolution of the global bathymetry data is about 2 minutes along latitudes and longitudes. Linear regression analysis reveals that the spatial accuracy ranges in the study area from 3.3 km at latitude 22° N to 3.1 km at latitude 28° N. That means the data are only adequate for recognizing features that have an extent of ≥ 3.1 km. Although Kriging interpolation procedure was used to create a smooth surface of cell size 100 m, the procedure does not increase accuracy of the foot print of the original data. However, valuable information has been obtained that manifest the main characteristics of the major submarine forms. Extracting smaller-scale features (e.g. estuaries, coral reefs, and submerged shallow water-courses) may need another data source (e.g. detailed multi-beam or echo-sounding survey) or enhanced digital image processing techniques that can be applied to high-resolution satellite images. The later field is open for more innovative approaches to estimate bathymetry from optical remote sensing data on the basis of numerical models and algorithms. The Egyptian Red Sea Coast in specific deserves further researches to manifest the characteristics of the micro-submerged coastal landforms.

The main objectives of this study are to recognize and clarify the major submerged and/or submarine landforms that marked the Egyptian Red Sea

Coast and the continental shelf. In that context, the study revealed the main coral reef types marked the coastal zone are namely fringing, barrier, atoll, patch, and carpet reefs. Fringing, barrier and patch reefs are generally common. Atolls are neither well developed nor spread. Carpet reefs are revealed in Hurghada by Riegl and piller (1997).

However, it is obvious that the continental shelf is noticeably influenced by the surface fault system extending eastward into the main Red Sea depression, showing the continental edge mostly like a fault-scarp of ~60° anticlockwise fault plane of steep slope. Sea ridges and subbasins were distinguished at the lower toe of the continental slope, which seem to be a result of the regional tectonic system. It is believed that the main fault and fold system belongs to the initial arching and rifting of the Red Sea (Tewfik and Ayyad, 1982; Said, 1990). Two sea peaks specifically were identified as extinct volcanoes owing to their abrupt rising from the sea floor and ideal conical shape with very steep sides. They are possibly attributed to the westward migration of the Red Sea asthenosphere during middle Miocene.

The study deduced as well that the submarine canyons are the most widespread features that break the continental edge. Two types of submarine canyons were recognized as deep incised canyons and shallow canyons. The deep incised canyons carve the continental edge with remarkable steep walls that seem to have been cut in the bedrock down to depth ~500 m bsl. This is in the agreement with the results discovered on the floor of the Red Sea by the U.K. Royal Navy's HMS Enterprise using an echo sounder survey. They might be formed as a result of the Messinian event (~5.59 Ma) owing to the sudden disconnection between the Atlantic Ocean and the Mediterranean during late Miocene times. Additionally, it is estimated that the shallow canyons are mostly developed during the Pleistocene lower sea level (some 90-130 m bsl) where the major wadis cut their water courses through the continental shelf. Some individual submerged deltas were identified based on the configuration of the bathymetric lines and their location at the topographic breaks. A close relationship has been found between the present-day drainage system and the deltas. It is inferred that the present-day drainage system is mostly a successor to an ancestor drainage system that was mostly developed during middle Miocene as the climate was turned toward warm humid (Khedr, 1984) or pre-middle Miocene as the climate was shifted toward a Sudano-Guinean type of savannah since the Oligocene (Le Houérou, 1997).

During Pleistocene and Holocene times the Red Sea experienced dramatic eustatic changes resulting in a sequence of raised coral terraces and alluvial terraces. Three main submarine terraces were identified at depths 20-25, 50-

75, 100-120 m bsl. Generally, they are in the agreement with the generalized curve of sea level rise since the LGM (~23-18 ka bp). Hence, it is deduced that terrace 100-120 m bsl is equivalent to LGM, terrace 50-75 m bsl ranges in age between 11.5 and 11 ka bp and terrace 20-25 m bsl is mostly formed during the Flandrian Transgression (8.2-7.6 ka bp). The Red Sea attained its present-day level mostly during the last 5-4 ka bp and the formation of the present-day estuaries may also in accordance with this period. It is expected that results of this study will be helpful for shedding light on the geomorphologic evolution of the continental shelf and its surrounding landmass.

ACKNOWLEDGMENTS

The author is deeply indebted to the anonymous reviewers for their valuable comments and suggestions to improve the quality of the manuscript.

REFERENCES

Ahmed, S.S., 1972. Geology and petroleum prospects in eastern Red Sea. *Am. Assoc. Petrol. Geol. Bull.*, Vol. 56, Issue 4, pp. 707 – 719.

Allan, T.D., 1970. Magnetic and gravity fields over the Red Sea. *Philos. Trans. R. Sot. London, Ser. A.*:270: 153-180.

Amante, C., Eakins, B.W., 2009. ETOPO1 1 Arc-Minute Global Relief Model: Procedures, Data Sources and Analysis. NOAA Technical Memorandum NESDIS NGDC-24, 19 p.

Arvidson, R., Becker, R., Shanabrook, A., Luo, W., Sturchio, N., Sultan, M., Lotfy, Z., Mahmood, A., El Alfy, Z., 1994. Climatic, eustatic and tectonic controls on Quaternary deposits and landforms, Red Sea, Egypt. *Journal of Geophysical Research*, Vol. 99, No. B6, pp. 12175-12190.

Bäcker. H., Lange, K., Richter, H., 1975. Morphology of the Red Sea central graben between Subair Islands and Abul Kizaan. *Geol. Jah*rb., D13: pp. 79-123.

Behairy, A. K. A., Sheppard, C.R.C., El-Sayed, M.K., 1992. A Review of the Geology of Coral Reefs in the Red sea. UNEP Regional Seas Reports and Studies.

Bonatti, E., Clocchiatti, R., Colantoni, P., Gelmini, R., Marinelli, G., Ottonello, G., Santacroce, R., Taviani, M., Abdel-Meguid, A., Assaf, S.,

El Tahir, M.A., 1983. Zabargad (St. John's) Island: an uplifted fragment of sub-Red Sea lithosphere. *J. geol. Soc. London*, Vol. 140, pp. 677-690.

Braithwaite, C.J.R., 1987. Geology and Paleogeography of the Red Sea Region. In: Edwards, A.J., Head, S.M. (Eds.), Red Sea. Pergamon Press, Oxford, pp. 22-44.

Chang, K. T., 2006. Introduction to Geographic Information Systems. 3rd ed. McGraw Hill, Singapore, 343 p.

Coleman, R.G., 1974. Geological background of the Red Sea. Deep Sea Drilling Project, Part III: Red Sea, doi:10.2973/dsdp. *Proc.* 23.121, pp. 813-819.

Daly , R.A., 1936. Origin of submarine canyons. *Am. J. Sci., Ser.* 5, 31, pp. 401-420.

Drake, C.L., Girdler, R.W., 1964. A geophysical study of the Red Sea. Geophys. *J. R. Astron. Sot.*, 8: pp. 473-495.

Edwards, A. J., Head, S.M. (Eds.), 1987. Red Sea. Pergamon Press, Oxford, 441 p.

El Moursi, M., Hoang, C. T., El Fayoumy, I. F., Hegab, O., Faure, H., 1994. Pleistocene evolution of the Red Sea coastal plain, Egypt: Evidence from uranium-series dating of emerged reef terraces. *Quaternary Science Reviews*, Vol. 13, pp. 345-359.

Embabi, N.S., 2004. The Geomorphology of Egypt: Landforms and Evolution. Vol. 1, *The Egyptian Geographical Society*, Cairo, 447p.

Emery, K.O., 1963. Sediments of the Gulf of Aqaba (Elat), R.L. Miller, Editor, Papers in Marine Geology, Macmillan, New York, pp. 257–273.

Fricke, H.W., 1983. On the origin of Red Sea submarine canyons. *Naturwissenschaften*, 70, pp. 195-197.

Gasse, F., 2000. Hydrological change in the African tropics since the Last Glacial Maximum, *Quaternary Science Reviews* 19, 189–212.

Gillmann, M., 1968. Preliminary results of a geological and geophysical reconnaissance of the Gizan coastal plain in Saudi Arabia. AIME Symp. Khahran, Saudi Arabia, pp. 198-208.

Gornitz, V., 2007. Sea level, after the ice melted and today. http://www.giss.nasa.gov/re-search/briefs/gornitz_09/ (accessed 17 July 2012).

Goudie, A.S., 2004. Deglaciation. In: Goudie, A.S., (Ed.), Encyclopedia of Geomorphology, vol.1, Routledge Ltd, London and New York, pp. 235-237.

Guennoc, P., Pautot, G., Coutelle, A., 1988. Surficial structures of the northern Red Sea axial valley from 23° N to 28° N: time and space evolution of neo-oceanic structures. In: Le Pichon, X., Cochran, J.R. (Eds.), The Gulf of Suez and Red Sea Rifting. *Tectonophysics,* 153: 1-23.

Guilcher, A., 1988. A heretofore neglected type of coral reef: the ridge reef, Morphology and origin. Online Summary available at: http://www.reefbase.org/resourcecenter/pub-lication/icrs.aspx (accessed 13. Juni. 2007).

Gvirtzman G., Buchbinder, B., Sneh, A., Nir, Y., Friedman, G.M., 1977. Morphology of the Red Sea fringing reefs: a result of the erosional pattern of the last-glacial low-stand sea level and the following Holocene recolonization. *Memoires Bureau Recherches Geologiques Minieres*, 89, pp. 480-491.

Gvirtzman, G., 1994. Fluctuations of sea level during the past 400.000 years: the record of Sinai, Egypt (northern Red Sea). *Coral Reefs*, vol. 13, pp. 203-214.

Hall, S.A., 1977. A total intensity magnetic anomaly map of the Red Sea and its interpretation. Ph.D. thesis, University of Newcastle Upon Tyne, 260 p.

Head, S. M., 1987. Corals and coral reefs of the Red Sea. In: Edwards, A. J. & Head, S. M. (ed.) Red Sea. pp. 128-151, Pergamon Press, Oxford.

Hoang, C.T., Taviani, M., 1991. Stratigraphic and tectonic implications of uranium series-dated coral reefs from uplifted Red Sea islands. *Quaternary Research*, Vol. 35, Issue 2, pp. 264-273.

Jokela, A.W., 1963. Submarine geology of the Red Sea. MSc. Department of geology and geophysics, *Massachusetts Institute of Technology*. 65p.

Jones, D. A., Ghamrawy, M., Wahbeh, M. I., 1987. Littoral and shallow subtidal environments. In: Edwards, A. J. & Head, S. M. (Ed.) Red Sea. pp. 169-193, Pergamon Press, Oxford.

Khedr, E.S., 1984. Sedimentological evolution of the Red Sea continental margin of Egypt and its relationship to sea-level changes. *Sedimentary Geology*, 39 pp. 71-86.

Knott, S.T., Bunce, E.T., Chase, R.L., 1966. Red Sea seismic reflection studies. In: Irvine, T.N. (Ed), The World Rift System. *Can. Geol. Surv., Pap.* 66-14, pp. 33-61.

Kolla, V., Macurda, Jr., 1988. Sea-level changes and timing of turbidity current events in deep-sea fan systems. In: Wilgus, C.K., Hastings, B.S., Kendall, C., Posamentier, H.W., Ross, C.A., Van Wagoner, J.C. (Eds.), Sea-level Changes, an international approach. Tulsa, Oklahoma, USA, *SEPM Special Publication* 42, pp. 381-392.

Krijgsman, W., Hiigeni, F.J., Raffi, I., Sierro, F.J., Wilson, D.S., 1999. Chronology, causes and progression of the Messinian salinity crisis. Nature, 400, pp. 652–655.

Laughton, A.S., 1970. A new bathymetric chart of the Red Sea. *Phil. Trans. Roy. Soc. London. A.* 267, pp. 21-22.

Laursen, J., Normark, W.R., 2002. Late Quatel1lary evolution of the San Antonio Submarine Canyon in the central Chile forearc. *Marine Geology* 188: pp. 365-390.

Le Houérou, H. N., 1997. Climate, flora and fauna changes in the Sahara over the past 500 million years. *Journal of Arid Environments*, Vol. 37, pp. 619-647.

Lloyd, C.D., 2007. Local model for spatial analysis. CRC Press (Taylor & Francis Group), USA, 260 p.

Maritime Reporter and Marine News Magazines Online, 2004. Uncharted Red Sea Underwater Mountain Discovered. http://www.marinelink.com/news/ underwater-uncharted346255.aspx (accessed 28 February 2013).

Marshak, S., Bonatti, B., Brueckner, H., Paulsen, T., 1992. Fracture-zone tectonics at Zabargad Island, Red Sea (Egypt). *Tectonophysics,* 216, pp. 379-385.

Moawad, B.M., 2008. Applications of remote sensing and geographic information systems in geomorphological studies: Safaga-El Quseir area, Red Sea, Egypt as an example. VDM Verlag Dr. Müller, Saarbrücken, 282 p.

Moawad, M.B., 2013. Detection of the submerged topography along the Egyptian Red Sea Coast using bathymetry and GIS-based analysis, Egypt. *J. Remote Sensing Space Sci.*, http://dx.doi.org/10.1016/ j.ejrs. 2012.12.001.

Nasr, A. H., 1955. Ecological observations on the marine algae of Ghardaqa (Red Sea). Hydrobiologia, Springer Netherlands, Vol. 7. No. 3, pp. 245-255.

National research Council, 1996. Alluvial fan flooding. National Academy Press, Washington. D.C., 172p.

Newman, A. C., McGill, D. A., 1962. Circulation of the Red Sea in early summer. Deep-Sea Res., Vol. 8: pp. 223-235. Pergamon Press Ltd. Great Britain.

Nima, 2000. Performance Specification, Digital Terrain Elevation Data (DTED®). MIL-PRF-89020B, 23 May.

O'Callaghan, J. F., and Mark, D. M., 1984. The extraction of drainage networks from digital elevation data. Computer Vision, *Graphics and Image Processing* 28: 323-44.

Orszag-Sperber, F., Plaziat, J-C., Baltzer, F., Purser, B.H., 2001. Gypsum salina-coral reef relationships during the Last Interglacial (Marine Isotope Stage 5e) on the Egyptian Red Sea coast: a Quaternary analogue for Neogene marginal evaporates? *Sedimentary Geology*, Vol. 140, pp. 61-85.

Pätzold, J., 2000. Report and preliminary results of Meteor Cruise M44/3, Aqaba (Jordan) - Safaga (Egypt) - Duba (Saudi Arabaia) - Suez (Egypt) – Haifa (Israel), March 12 - March 26 - April 2 - April 4, 1999. Berichte aus dem Fachbereich-5 Geowissenschaften der Universität Bremen, No. 149, 135p.

Pautot, G., Guermoc, P., Coutelle, A., Lyberis, N., 1986. La dèpression axiale du segment nord mer Rouge nouvelles donèes gèologiques et gèophysiques obtenues au cours de la campagne TRANSMEROU 83. *Bull. Soc. Gèol. Fr.*, 8: 381-399.

Phillips, J.D., Ross, D.A., 1970. Continuous seismic reflection profiles in the *Red Sea. Philos. Trans. R. Sot. London, Ser. A*. 267: pp. 205-217.

Plaziat, J-C., Baltzer, F., Choukri, A., Conchon, O., Freytet, P., Orszag-Sperber, F., Purser, B., Raguideau, A., Reyss, J-L., 1995. Quaternary changes in the Egyptian shoreline of the northwestern Red Sea and Gulf of Suez. *Quaternary International*, Vol. 29/30, pp. 11-22.

Por, F. D., 1972. Hydrological notes on the high-salinity waters of the Sinai Peninsula. *Marine Biology*, Vol. 14, pp. 111-119.

Radwan, G. El-D. M., 2001. Impact of human activities on coral reef sedimentary environment in Mabahiss bay, north to Hurghada city – Red Sea. *Unpub. M. Sc. Thesis. Geol. Dept. Fac. Sci.* Cairo University, 132 p.

Reaches, Z., Erez, J., Garfunkel, Z., 1987. Sedimentary and tectonic features in the north eastern Gulf of Elat, Israel. *Tectonophysics,* Vol. 141, pp. 169-180.

Reiss, Z., Hottinger, L., 1984. The Gulf of Aqaba: ecological micropaleontology. Ecological Studies, vol. 50. Springer Verlag, Berlin, 354 pp.

Riegl, B., Piller, W. E.,1999. Coral frameworks revisited-reefs and coral carpets in the northern Red Sea. *Coral Reefs,* 18 : 241-253.

Said, R., 1990. Red Sea Coastal Plain. In Said, R., (Ed.) The geology of Egypt. A. A. Balkema Rotterdam, pp. 345-359.

Shepard, F. P., Dill, R.F., 1966. Submarine canyons and other Sea Valleys. Rand Mcnally, Chicago, 381 pp.

Shepard, F.P., 1972. Submarine canyons. Earth-Science Reviwes, Elsevier, Amesterdam. 12p.

Shukri, N. M., Higazy, R.A., 1944. Mechanical analysis of some bottom deposits of the northern Red Sea. Jour. *Sedimentary Petrology*, 14, pp. 43-69.

Smith, W.H.F., Sandwell, D., 1997. Global seafloor topography from satellite altimetry and ship depth soundings. *Science* 277, 1956–1962.

Stoffers, P., Ross, D.A., 1974. Sedimentary history of the Red Sea. *Deep Sea Drilling Project,* Part III: Red Sea, doi:10.2973/dsdp. Proc. 23.125, pp. 849-865.

Stow, D.A.V., Howell, D.G., Nelson, C.A., 1985. Sedimentary, tectonic, and sea-level controls. In: Bouma, A.H., Normark, W.R., Barnes, N.A. (Eds.), Submarine fans and related turbidity systems. Springer-Verlag, New York, pp. 15-22.

Strahler, A. N., 1957. Quantitative analysis of watershed geomorphology. *Am. Geophys. Union. Trans*. 38 (6), pp. 913-920.

Tarboton D. G., Bras, R. L., Rodriguez-Iturbe, I., 1991. On the extraction of channel networks from digital elevation data. *Hydrological Processes,* Vol. 5, pp. 81-100.

Tewfik, N., Ayyad, M., 1982. Petroleum exploration in the Red Sea shelf of Egypt. E.G.P.C., six exploration seminars, Cairo.

Thompson, A., 2013. Huge Canyon Discovered Beneath Red Sea. http://www.livescience.com/27498-red-sea-canyon-discovered.html (accessed 28 February 2013).

Tomczak, M., 1996-2000. Oceanogr. www.es.flinders.ed.au (accessed on 05 September 2007).

Uchupi, E., Ross, D.A., 1986. The Tectonic Style of the Northern Red Sea. *Geo-Marine Letters* 5: pp. 203-209.

U.K. Royal Navay, 2013. HMS Echo. http://www.royalnavy.mod.uk/The-Fleet/Ships/Patrol-and-Minehunters/Ice-Patrol-and-Survey-Ships/HMS-Echo (accessed 28 February 2013).

Unlimited Media, 2004. Nature's beauty: The Red Sea. Underwater video production. Dolby Digital and absolute records GmbH.

Vandersträte, T., 2006-2007. The use of remote sensing for coral reef mapping in support of integrated coastal zone management: a case study in the NW Red Sea. Ph.D. Universiteit of GENT Faculteit Wetenschappen, Vakgroep Geografie, Belgium. 426 p.

In: Continental Shelf
Editors: L. Scanlon and J. L. Ranieri ISBN: 978-1-62948-330-6

Chapter 3

BENTHIC BOUNDARY LAYER (BBL) MACROFAUNAL COMMUNITY STRUCTURE OVER OLIGOTROPHIC CONTINENTAL SHELVES: A NEW FIELD METHODOLOGICAL APPROACH

Panayota Koulouri*
Institute of Marine Biology, Biotechnology & Aquaculture, Hellenic Centre for Marine Research, Gournes Pediados, Heraklion, Crete, Greece

ABSTRACT

The marine sediment-water interface is one of the most extensive habitats on the planet occupying 75% of the Earth's surface. Fluxes of materials across the sediment-water interface and the mechanisms that mediate and constrain those fluxes are likely to have global importance. Because of differing ecosystem structure above and below the sediment-water interface, the ecologists who study these domains use different techniques and ask different research questions leading often to scientific isolation. Consequently, the significance of the fauna related to the sediment-water interface is underestimated, especially in studies of benthic-pelagic coupling related to energy fluxes and also in holistic approaches of the marine ecosystem. The portion of the sediment and the

* E-mail: yol72@hcmr.gr.

water column that is influenced by the sediment-water interface is characterized by strong gradients of flow as well as dissolved and particulate matter. This dynamic zone, also known as benthic boundary layer (BBL), can be extended from a few centimetres to tens of metres, depending both temporally and spatially on various physical and biogeochemical processes whereby organisms on either side of the sediment-water interface interact. BBL supports both benthic and pelagic, mostly macrofaunal, organisms classified in different groups according to their different mobility and bottom dependence. BBL species (mainly crustaceans) are considered to be key taxa in marine food webs as most of them exploit a large diversity of food resources, mobilizing organic and detritus particles and associated microbes or preying on meiobenthos and plankton, while at the same time constituting an important food source for demersal fish, epibenthic crustaceans and cephalopods, of high commercial value. For the study of BBL macrofauna, specially designed samplers, generally known as hyperbenthic or suprabenthic sledges, have been used over the last 40 years. Nevertheless, there are still *practical* technical difficulties in sampling efficiently the lowest layer of the water column above the seabed where most of these organisms are concentrated. In the framework of an EU-funded project, a new sampling device for the study of BBL macrofaunal communities was developed and applied experimentally in the field. This novel sampling technique included, as a complementary application, a specially designed facility attached in the frontal part of a modified hyperbenthic sledge for the artificial resuspension of the surface sediment. This facility provided additional quantitative information on the BBL macrofaunal populations concentrated in the lowest layer of the water column, just above the seabed surface, allowing a more comprehensive view of the role of this particular macrofauna in the marine food web to be built. The application of this novel sampling approach gave new insights towards a better understanding of the role of biodiversity related to the sediment-water interface with respect to nutrient regeneration, carbon cycling and energy transfer to higher trophic levels as well as to relevant bottom trawling effects.

Introduction

The Benthic Boundary Layer (BBL) Macrofauna

The Benthic Boundary Layer (BBL), which could be considered as the portion of the sediment and the water column that is influenced by the sediment-water interface, is characterized by strong gradients of flow as well as dissolved and particulate matter, i.e. lower water movement and higher concentrations of particles and biomass of organisms than in the water layer immediately above it (Boudreau & Jørgensen, 2001). This dynamic zone can be extended from a few centimetres to tens of metres, e.g. 10 cm to 5m in coastal habitats, depending both temporally and spatially on various physical and biogeochemical processes whereby organisms on either side of the sediment-water interface interact (Marshall & Merret, 1977). BBL supports epibenthic, hyperbenthic and pelagic macrofaunal organisms classified in different groups considering their different mobility and bottom dependence. According to Dauvin & Vallet (2006), these groups are: (i) permanent (e.g. copepods, chaetognaths, euphausiids) and temporary (e.g. crustacean and fish larvae) swimmers present only in the BBL or throughout the water column, which are also considered as zooplankton; (ii) burrowers/swimmers, which correspond to hyperbenthos including all bottom dependent animals (mainly crustaceans) which perform - with varying amplitude, intensity and regularity - seasonal or daily vertical migrations above the seafloor (Brunel et al., 1978); (iii) burrowers and temporary nuptial swimmers emerging during the nuptial period (e.g. polychaete syllids and nereids) as well as benthic mobile and drift species (e.g. bivalves and polychaetes), most of which are epibenthic, and (iv) fish.

Consequently, there is little acknowledgment for BBL macrofauna partly attributed to the broad diverse forms related to distribution in space and not to phylogeny or functionality as well as to the use of different terms and definitions (Mees & Jones, 1997). The terms "hyperbenthos" or "suprabenthos" are used for this particular macrofauna mainly in the continental shelf of temperate and northern areas (e.g. Sorbe, 1989; Cunha et al., 1997a;b; Dewicke et al., 2002; 2003). The same fauna is also called as "demersal zooplankton", "resident" or "swarming" in tropical areas (Youngbluth, 1982; Carleton & Hamner, 1989; Kim & Oliver, 1989) and as "benthopelagic plankton" or "BBL macrofauna" in the deep sea (Wishner, 1980; Dauvin et al., 2000). Mees & Jones (1997) refer to many other terms which are used for the same faunal assemblage or for groups of these animals, e.g. "motile epibenthos", "hypoplankton", "micronekton", "nektobenthos", "near bottom zooplankton", etc.

However, BBL macrofauna is considered to be an important link in marine food webs (Mees & Jones, 1997). Most of these organisms exploit a large diversity of food resources, i.e. organic and detritus particles and associated microbes, meiobenthos, plankton (Fanelli et al., 2009a;b). These animals also constitute an important food source for demersal fish, and especially for their 0-group individuals, (Sorbe, 1981; Wildish et al., 1992; Martin & Christiansen, 1997; Labropoulou & Eleftheriou, 1997; Cartes et al., 2002), epibenthic crustaceans (Cartes 1993a;b;c; Cartes & Maynou, 1998; Labropoulou & Kostikas, 1999) and cephalopods (Bergstrom 1985; Villanueva & Guerra, 1991). Many of these predators are of high commercial value and therefore important for fisheries.

Methodological Approaches for the Study of BBL Macrofauna

The role of BBL macrofauna in the marine ecosystems is mostly underestimated due to the inaccessibility of its particular environment. Sampling these often highly mobile animals is not easy as they are caught occasionally by conventional benthic (e.g., grabs, corers) or pelagic (e.g. plankton nets) sampling gears (Eleftheriou, 2013). Specially designed samplers, generally known as hyperbenthic or suprabenthic sledges, have been designed and used over the last 40 years for quantitative or semi-quantitative sampling of the BBL macrofaunal communities at various distances from the soft-bottom seafloor (Macer, 1967; Hetshagen, 1973; Rothlisberg & Pearcy, 1977; Brunel et al., 1978; Sorbe, 1983; Dauvin & Lorgeré, 1989; Cartes et al., 1994a;b; Cunha et al., 1997a;b; Brandt & Schnack, 1999; Dauvin et al., 2000; Dewicke et al., 2002; 2003; Eleftheriou, 2013). However, most of these sledges sample at least 10 cm above the seabed because sampling very close to the bottom results in contamination of the samples with both sediment and organisms living in or on the sediment (Table 1). Semi-quantitative epibenthic sledges (e.g. WHOI, SOC, AQUAREVE III; Eleftheriou, 2013) as well as modified dredges and trawls (e.g. Triple D, 2-m modified beam trawl; Eleftheriou, 2013) are usually equipped with nets of large mesh size (> 5 or 10 mm) in order to avoid clogging with sediment (Table 1). Modified sledges with stimulation devices like tickler chains (Ockelmann, 1964) and water-jet or electric currents (Turnbull & Watson, 1992) have also been applied, but only sporadically and for qualitative sampling of certain animals found very close to the bottom and on or in the sediment. Consequently, there are still *practical* technical difficulties in sampling efficiently the lowest layer of the water column above the seabed, where most of these organisms are concentrated (Mees & Jones, 1997; Eleftheriou, 2013; Koulouri et al., 2013).

Table 1. Technical details of different samplers used for the study of BBL macrofauna

Type of sampler	No of nets	Level of sampling above the seabed (cm)	Mouth opening of the nets (width x height cm^2)	Measurement of the sampling surface or the water filtered	Mesh-size of the nets (mm)	Opening-closing mechanism of the nets	References
Beyer's sledge	1 or 2 or 3	0-80 or 0-130	50 cm (diameter)	flowmeter	0.5	mechanic or messenger	Hesthagen, 1973; Boysen, 1975; Oug, 1977; Hesthagen & Gjermundsen, 1978; Fossa, 1985; Beyer & Indrehus, 1995
WHOI Epibenthic Sled (Woods Hole Oceanographic Institution)	1	<0-30	81x30		0.5 or 1	electronic	Hessler & Sanders, 1967; Huberdeau & Brunel, 1982; Eleftheriou, 2013
Sorbe's suprabenthic sledge	2	0-100		odometer, flowmeter	0.5 or 1	mechanic	Sorbe, 1983; Sorbe, 1989; Cunha et al., 1997a;b; Dewicke et al., 2002; 2003
Macer-GIROQ suprabenthic sledge	2 or 3 or 4	28-142 or 33-149 or 10-145	46x38 or 80x40 or 60x30	odometer, flowmeter	0.5	mechanic	Macer, 1967; Brunel et al., 1978; Dauvin & Lorgeré, 1989; Wildish et al., 1992; Cartes et al., 1994a;b; Dauvin et al., 1995; 2000
Epibenthic sled AQUAREVE III	1 or more nets	(-2.5)-40	34x40	odometer	10	electro-mechanic	Thouzeau & Vine, 1991; Rowell et al., 1997; Eleftheriou, 2013

Table 1. (Continued)

Type of sampler	No of nets	Level of sampling above the seabed (cm)	Mouth opening of the nets (width x height cm^2)	Measurement of the sampling surface or the water filtered	Mesh-size of the nets (mm)	Opening-closing mechanism of the nets	References
Epibenthic sampler of RP	1 or 2	2.5-22.9 or 26-60 or 27-60 and 100-133	30x30	flowmeter, GPS	0.5 or 0.571	mechanic or pneumatic	Rothlisberg & Pearcy, 1977; Buhl-Jensen, 1986; Fossa et al., 1988; Brattegard & Fossa, 1991; Brandt & Barthel, 1995; Brandt & Berge, 2007
SOC Epibenthic Sledge (UK Institute of Oceanographic Sciences)	1 or 2 or 3	80 or 80-122	230x60 or 120x50	flowmeter, odometer	0.3 or 1and 4.5	acoustic or electro-mechanic	Aldred et al.. 1976; Rice et al., 1982; Christiansen & Nuppenau, 1997; Eleftheriou, 2013
Deep Digging Dredge (Triple-D)	1	(-10)-(-5)	150x150	odometer	>10		Bergman & Van Santbrink, 1994; Eleftheriou, 2013
2-m modified beam trawl	1	<0		Decca navigator	10 or 20		Kaiser et al., 1994; Eleftheriou, 2013
TTSS2 (Towed Trawl Simulator Sledge)	3	<0-83	58x26	odometer	0.5	electro-mechanic	Koulouri et al., 2003; Eleftheriou, 2013

Figure 1. The Towed Trawl Simulator Sledge v2 (TTSS2) for the study of BBL macrofauna (technical details are described in Koulouri et al., 2003).

Development of Innovative Sampling Techniques for the Study of BBL Macrofauna

In the framework of an EU-funded project aiming at developing a new method for the study of the direct effects of bottom trawling on benthic nutrient fluxes and sediment biogeochemistry (Dounas et al., 2002), a new sampling device, known as TTSS2, was developed and applied in the field in order to study BBL macrofaunal communities (Koulouri et al., 2003). This sampling technique includes, as a complementary application, a specially designed facility attached to the frontal part of a modified hyperbenthic sledge for the artificial resuspension of the surface sediment (Figure 1). The TTSS2, equipped with plankton nets (0.5 mm mesh size) and an opening-closing mechanism, quantitatively samples three levels of the water column (0.05-0.30 m, 0.31-0.56 m, and 0.57-0.82 m) above the seabed. The apparatus also includes an odometer for measuring the distance travelled by the gear as well as a continuously recording video camera. Deployment of the TTSS2 can be made with and without artificial resuspension of the surface sediment for quantitative sampling of BBL macrofaunal communities. Samples collected without artificial resuspension of the surface sediment can also be used as reference for those collected with artificial resuspension in order to estimate BBL macrofaunal organisms found very close to the seabed (0 - 5 cm). The TTSS2 is the second modified version around the frame of the towed benthic sledge described by Shand & Priestley (1999), while the first version (TTSS or TTSS1) is equipped with water sampling bottles (Figure 2) in order to investigate environmental parameters within the BBL habitat as well as physical and geochemical effects on surface sediments caused by trawling (Dounas, 2006; Dounas et al., 2005; 2007).

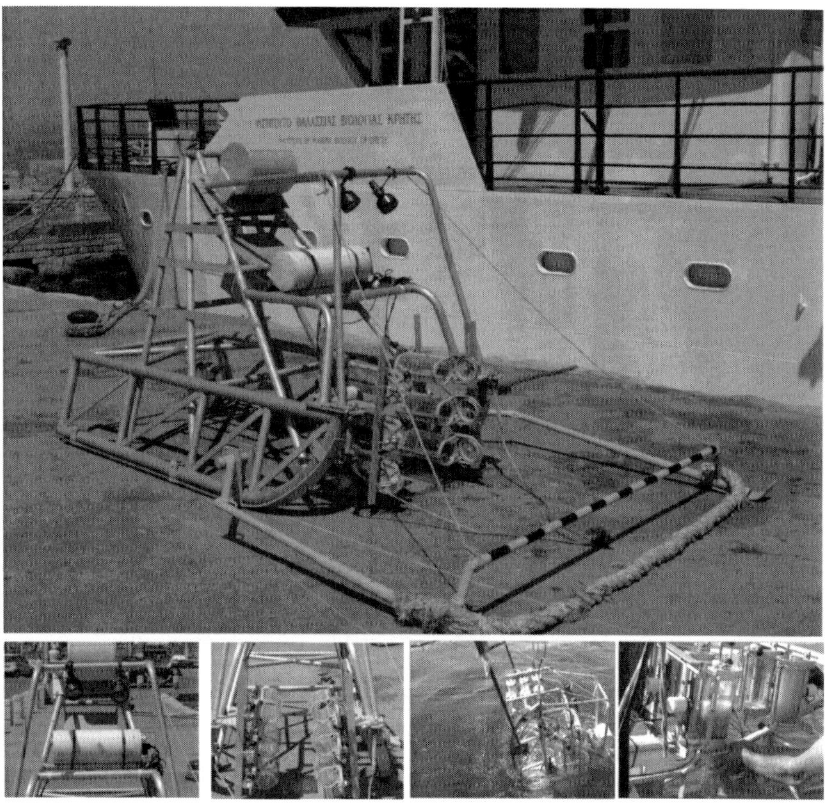

Figure 2. The Towed Trawl Simulator Sledge v1 (TTSS or TTSS1) for the study of BBL biogeochemistry (technical details are described in Dounas, 2006).

BBL Macrofaunal Communities in the Continental Shelf

There have been many studies investigating BBL macrofaunal components in many places around the world, and especially in temperate continental shelf seas. Several researchers used Beyer's sledge in order to study the hyperbenthos in the western Baltic Sea (Hesthagen, 1973; Boysen, 1975; Hesthagen & Gjermundsen, 1979) and Norwegian fjords (Oug, 1977; Hetshagen & Gjermundsen, 1978; Kaartvedt, 1985). A Rothlisberg & Pearcy (RP) sledge (1977) was also used in the Norwegian and Swedish shelf (Buhl-Jensen, 1986; Brategard & Fossa, 1991; Buhl-Jensen & Fossa, 1991). Hyperbenthic fauna of the North Sea have been investigated (Dewicke et al., 2002; 2003) by using the sledge described by Sorbe (1983). The BBL macrofaunal communities of the English Channel (eastern Atlantic) have been studied in detail and for many years with Macer-GIROQ sledge and its modified versions (e.g. Wang & Dauvin, 1994; Zouhiri & Dauvin, 1996; Dauvin & Zouhiri, 1996; Vallet & Dauvin, 1998; 1999; Dauvin et al., 1994; 2000). The suprabenthic soft-bottom communities of the south Gascogne (eastern Atlantic) continental shelf were also described (Cornet et al., 1983; Sorbe, 1989) by using Sorbe's sledge. Cunha et al., (1997a;b) investigated suprabenthos in the Portuguese continental shelf with the same sledge but with slight modifications. The structure of hyperbenthic communities in the Bay of Fundy, Baie des Chaleurs and the St. Lawrence Gulf (northwestern Atlantic) was extensively studied with Macer's sledge (Brunel et al, 1978; Sainte-Marie & Brunel, 1983; 1985; Chevrier et al., 1991). Wildish et al. (1992) used the same, but modified, gear to sample BBL macrofauna of Browns Bank (northwestern Atlantic) over hard bottoms. Lately, expeditions for investigation of the hyperbenthic fauna have taken place in Arctic and Antarctic areas by using modified RP and Macer-GIROC sledges, respectively (Brandt, 1995; Brandt & Barthel, 1995; Brandt & Schnack, 1999; Brandt & Berge, 2007; San Vicente et al., 1997; 2006; 2009). However, surveys conducted on the continental shelf of the western Mediterranean Sea are scarce, e.g. the Gulf of Marseille (Champalbert & Macquart-Moulin, 1970), the Ebro River Delta (Cartes et al., 2007), the northern Adriatic Sea and the northern Sicily coasts (Ligas et al., 2007; Fanelli et al., 2009a; 2011). The Macer-GIROQ sledge was used for most of these studies. As far as the eastern Mediterranean is concerned, BBL macrofauna has received little attention. The application of the TTSS2 in Heraklion Bay (Crete) is the first attempt to study BBL macrofaunal communities in the continental shelf of eastern Mediterranean (Koulouri et al., 2003; 2005; 2006; 2009; 2011; 2012; 2013).

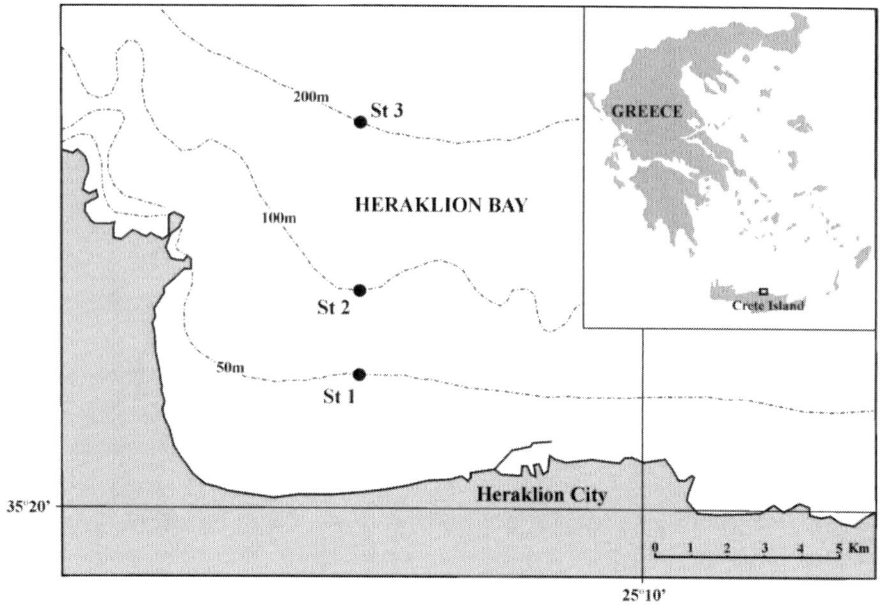

Figure 3. Map and sampling locations of the continental shelf of Heraklion Bay (Crete, E. Mediterranean).

Continental shelf of the Eastern Mediterranean: The Case of Heraklion Bay

The continental shelf of Heraklion Bay, located on the northern coast of Crete (Greece, E. Mediterranean), covers an area of 110 km^2 exposed to wave action and characterized mostly by muddy sediments (Figure 3). The coastal zone is affected by small-scale freshwater inputs and urban waste discharges (Karakassis & Eleftheriou, 1997), while the offshore system is characterised by marked oligotrophy, since the annual primary productivity has been approximately estimated to be 80gC/m^2/year by Psarra et al. (2000). Local environmental conditions are characterized by strong water mixing and relatively high pelagic primary productivity during late winter-early spring. On the other hand, water masses appear to be strongly stratified and with extremely low primary productivity in late summer-early autumn (Psarra et al., 2000). During this period the continental shelf seafloor receives enough light to support benthic primary productivity which seems to characterize the muddy sediments over a large part of the continental shelf of Heraklion Bay (Dounas et al., 2002; 2007). During late winter-early spring, deep chlorophyll a maxima (DCM) have been consistently observed between 75 and 120 m in the shelf of Heraklion Bay resulting in relatively high primary production in the water column (Psarra et al., 2000; Tselepides et al., 2000), which probably leads to accumulated algal matter in the near bottom environment (Dounas et al., 2002; 2007).

FIELD EXPERIMENTS

During the last ten years, different sets of field experiments have been carried out in the field and the application of the TTSS2 equipped with the artificial resuspension facility has provided additional quantitative information on the BBL macrofaunal communities concentrated in the lowest layer of the water column, just above the seabed. This information has led to a more comprehensive view of the role of this particular fauna in the oligotrophic continental shelf of the eastern Mediterranean (Koulouri et al., 2003; 2009; 2013). The artificial resuspension facility has also been applied in order to investigate the direct effects of bottom trawling on BBL macrofaunal communities which are highly impacted due to this anthropogenic pressure (Koulouri et al., 2003; 2005). The various sets of field experiments, including the application of the sampling technique equipped with the artificial resuspension facility (TTSS2), were planned in order to: (1) study the BBL macrofaunal component in comparison with the macrofauna collected by using other traditional benthic, hyperbenthic and planktonic samplers; (2) describe the qualitative and quantitative structure of BBL macrofaunal communities considering bathymetric, seasonal and diel changes as well as the prevailed environmental conditions; (3) investigate the effects of otter trawling on BBL macrofauna.

Application of the TTSS2 and Comparison with Traditional Benthic, Hyperbenthic and Planktonic Samplers

The first set of experiments involved the comparison of samples collected by using the TTSS2 as well as traditional benthic (Smith McIntyre grab), hyperbenthic (hyperbenthic sledge, HS) and planktonic (WP2 plankton net) samplers at the mid-shelf of Heraklion Bay (Koulouri et al., 2003). The differences between the sampling gears used have shown the relative efficiency of each sampler in collecting organisms characterized by different behaviour and distribution patterns (Figure 4). Polychaetes are numerically dominant in the macro-infaunal community collected with Smith McIntyre grab, followed by sipunculans, decapods and bivalves. The application of the TTSS2 has revealed a rich BBL macrofaunal community dominated by hyperbenthic crustaceans (e.g. mysids, amphipods, cumaceans) and epibenthic gastropods. In contrast, the activity of these organisms above the seabed seems to be extremely limited when the typical HS is used due to their close association with the sediment, at least during daylight hours (Koulouri et al., 2003; 2013). Moreover, patterns of taxa composition in the samples of HS appear to be similar with the one found in the vertical hauls of the WP2 plankton net, with a high proportion of crustaceans (mostly crustacean larvae and calanoid copepods) followed by cnidarians and chaetognaths. The WP2 plankton net may sample zooplankton adequately but is not really useful for collecting the fauna concentrated temporarily or permanently in the water column close to the seabed.

As it is widely accepted that BBL macrofaunal diversity comprises a mixture of taxa with different ecological habits and distribution patterns near the bottom, complementary information derived from different sampling procedures remains essential. However, the macrofauna collected by using the TTSS2 equipped with the artificial sediment resuspension facility includes small-sized epibenthic, hyperbenthic and zooplanktonic animals living or sheltering in the lowest part of the BBL. Furthermore, stomach content analysis of local bentho-pelagic fish species has shown that prey composition and abundance are strongly reflected in BBL macrofaunal organisms collected by using the TTSS2, which therefore can be considered as an effective tool for describing trophic interactions within the BBL habitat (Koulouri et al., 2011).

BBL Macrofaunal Community Structure and Bathymetric Distribution

The second set of experiments aimed at studying the BBL macrofaunal communities in the continental shelf of Heraklion Bay (Koulouri et al., 2011). The application of the TTSS2 modified to resuspend the surface sediment has given a more realistic estimation of the BBL macrofauna, at least in the oligotrophic environment of the study area, than the typical HS because the largest proportion of this particular fauna is found residing within a few centimetres (0-0.05 m) above the seafloor. BBL macrofaunal densities collected with artificial resuspension of the surface sediment are up to three orders of magnitude higher than those estimated in the samples collected without any sediment resuspension (Figure 5A). Diversity (especially for the benthic species of polychaetes, molluscs and crustaceans) is also much higher, despite the 3 to 10-fold length of tows performed by the HS (Figure 5B).

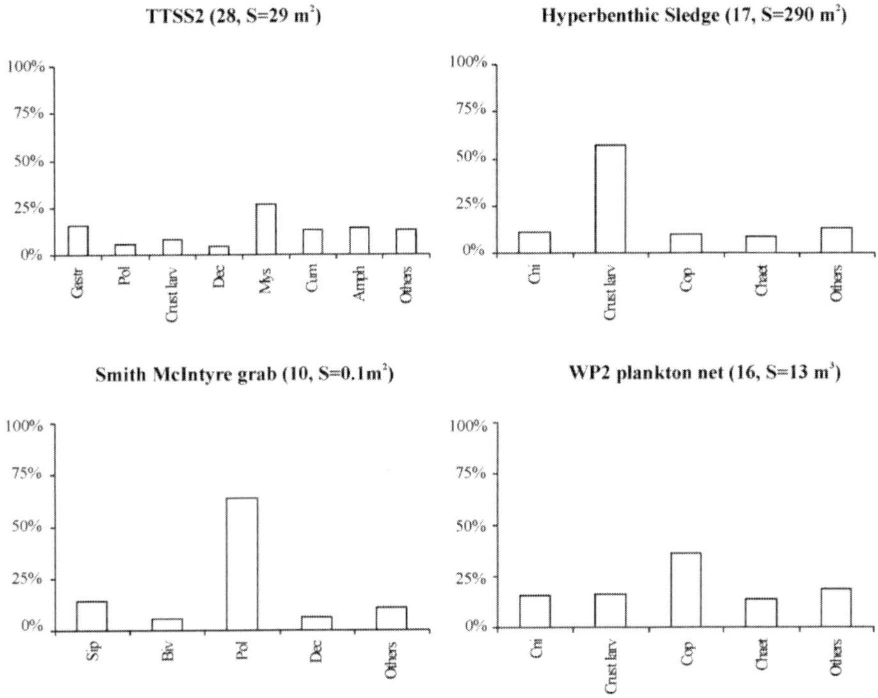

Figure 4. Percentages of the macrofaunal groups (Gastr: Gastropoda, Pol: Polychaeta, Crust larv: Crustacean larvae, Dec: Decapoda, Mys: Mysidacea, Cum: Cumacea, Amph: Amphipoda, Cni: Cnidaria, Cop: Copepoda, Chaet: Chaetognatha, Sip: Sipuncula, Biv: Bivalvia) collected with the four different samplers at the mid-shelf of Heraklion Bay (50 m). The number of the macrofaunal groups collected by each sampler and the sampling surface (S) are indicated in parentheses.

The BBL macrofauna along the different depths of the continental shelf of Heraklion Bay is characterized mainly by crustaceans, such as the cumaceans *Iphinoe rhodaniensis, Diastyloides serratus, Campylaspis sulcata, Campylaspis glabra,* the mysids *Anchialina agilis, Haplostylus lobatus, Erythrops* cf. *peterdohrni, Hypererythrops* sp., the amphipods *Perioculodes longimanus longimanus, Rhachotropis integricauda* and the decapod *Philocheras bispinosus* (Koulouri et al., 2013). The hyperbenthic crustacean diversity of Heraklion Bay is relatively rich and similar in structure to others described in the continental shelf of Mediterranean and European areas, while densities seem to be relatively lower and influenced mostly by the availability of food resources (Koulouri et al., 2013). However, near-bottom zooplanktonic densities recorded (e.g. crustacean larvae, chaetognaths and copepods) are similar to those in other eutrophic areas and probably form a considerable fraction of the zooplankton populations living throughout the water column (Koulouri et al., 2009).

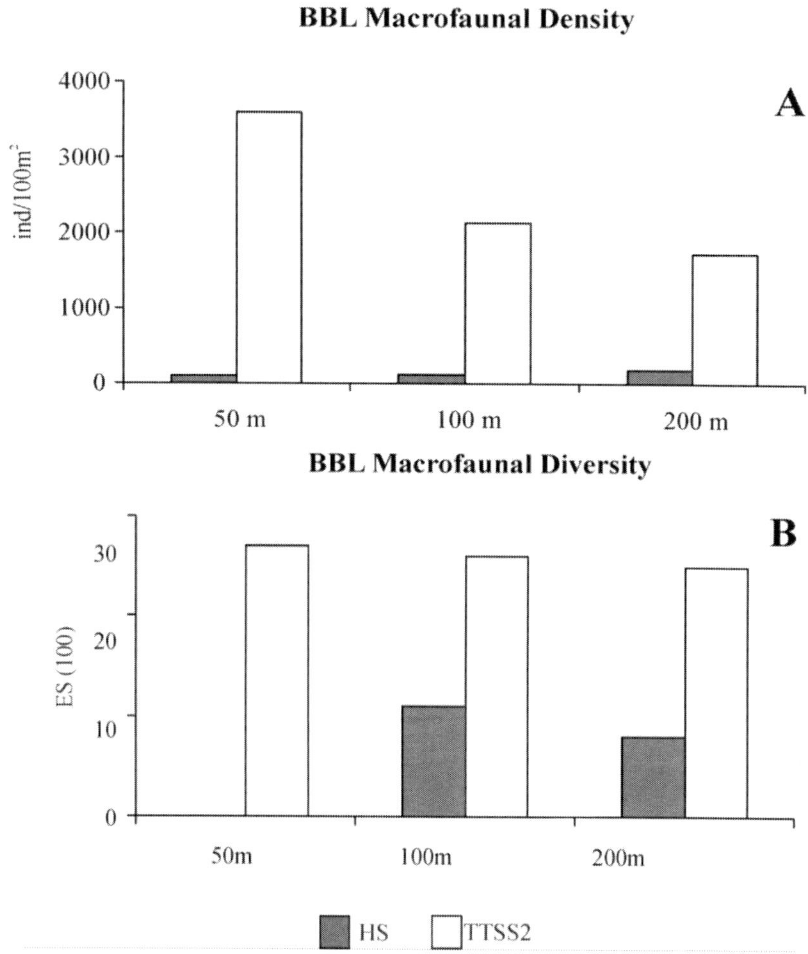

Figure 5. BBL macrofaunal density (A) and diversity (B) collected with TTSS2 (S=29-116 m^2) and HS (S=290 m^2) over the continental shelf of Heraklion Bay (S: sampling surface; ES(100): expected number of species for a theoretical sample of 100 specimens).

Seasonal Changes

The second set of experiments also included sampling for the investigation of seasonal changes of the hyperbenthic fauna in the continental shelf of Heraklion Bay (Koulouri et al., 2013). The seasonal bathymetric distribution of hyperbenthos seems to be influenced by the availability of food sources in the study area. The extremely high densities of hyperbenthos residing at the sediment-water interface in early autumn (Figure 6A) are probably related to the indicative presence of relatively high benthic primary productivity during this period in the study area. Most of the hyperbenthic species remain on or just a few centimetres above the seabed during early spring (despite the fact that it looks as if a very small part of this particular fauna moves high up above the bottom (Figure 6B)). The increased amounts of high quality organic matter (pelagic primary productivity) along with low levels of illumination very close to the seabed may explain the slight swimming activity near the seabed of a few individuals during this period. The stronger inputs of organic matter in the shallow part of the continental shelf also seem to be the driving force for seasonal horizontal migrations of certain hyperbenthic species.

Diel Changes

The same set of experiments also included sampling during the night for the investigation of diel changes of hyperbenthic and zooplanktonic animals in the different layers of the water column at the mid-shelf of Heraklion Bay (Koulouri et al., 2009; 2013). Significant changes in abundances near the bottom environment during the night are indicative of nocturnal vertical migrations performed by certain taxa (Figure 7). Most of the hyperbenthic amphipods, cumaceans, decapods and mysids seem to shelter in or on the sediment surface during the day, probably in order to avoid predators, and most of these emerge or swim up to the near-bottom water layer during the night. The limited amplitude of these migrations in the oligotrophic system of the study area is probably a response to the relatively high concentrations of labile organic matter in the sediment surface (Koulouri et al., 2013). Furthermore, near-bottom zooplankton (e.g. crustacean larvae, chaetognaths, calanoid copepods) seems to exhibit "reverse" diel vertical migrations at the same area, being also highly concentrated at the sediment-water interface during the night (Koulouri et al., 2009).

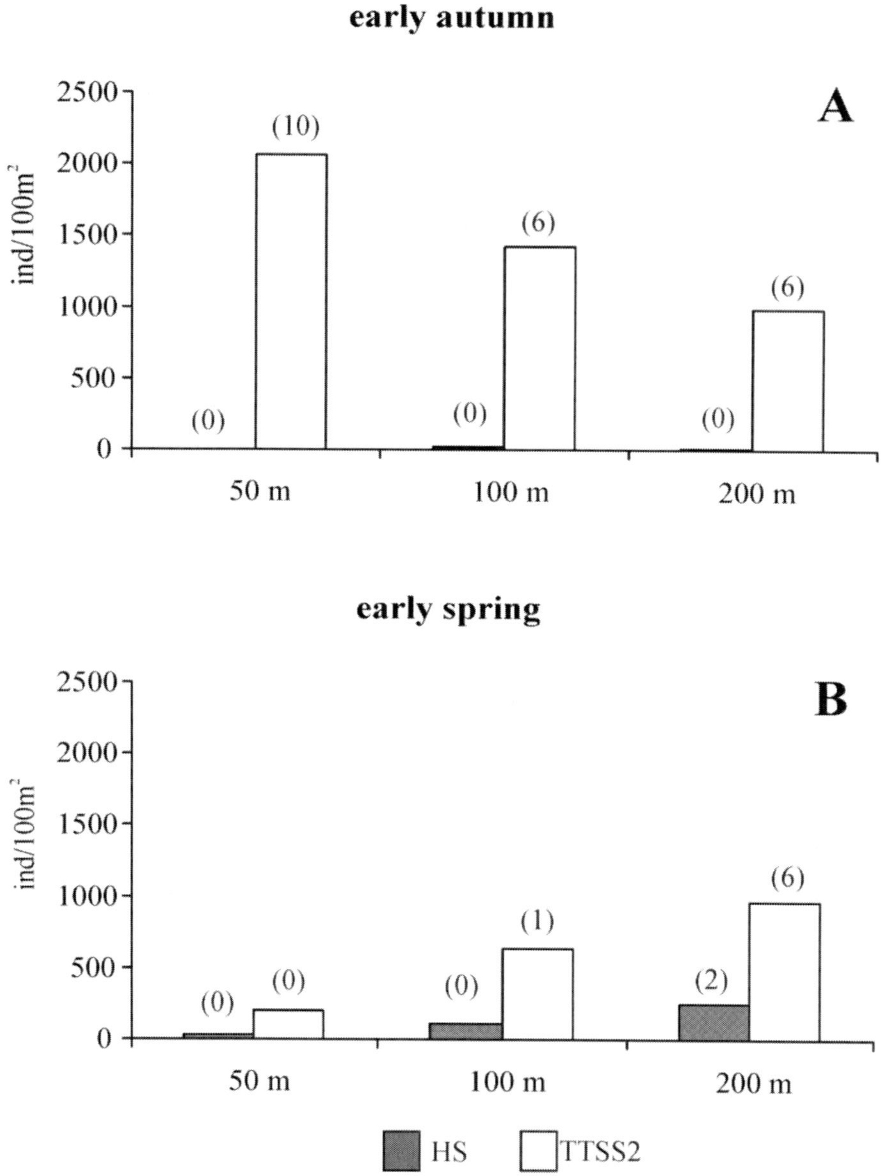

Figure 6. Bathymetric distribution of the BBL macrofauna collected with TTSS2 and HS in the continental shelf of Heraklion Bay during two seasonal occasions. Expected number of species for a theoretical sample of 100 specimens, ES (100), is indicated in parentheses.

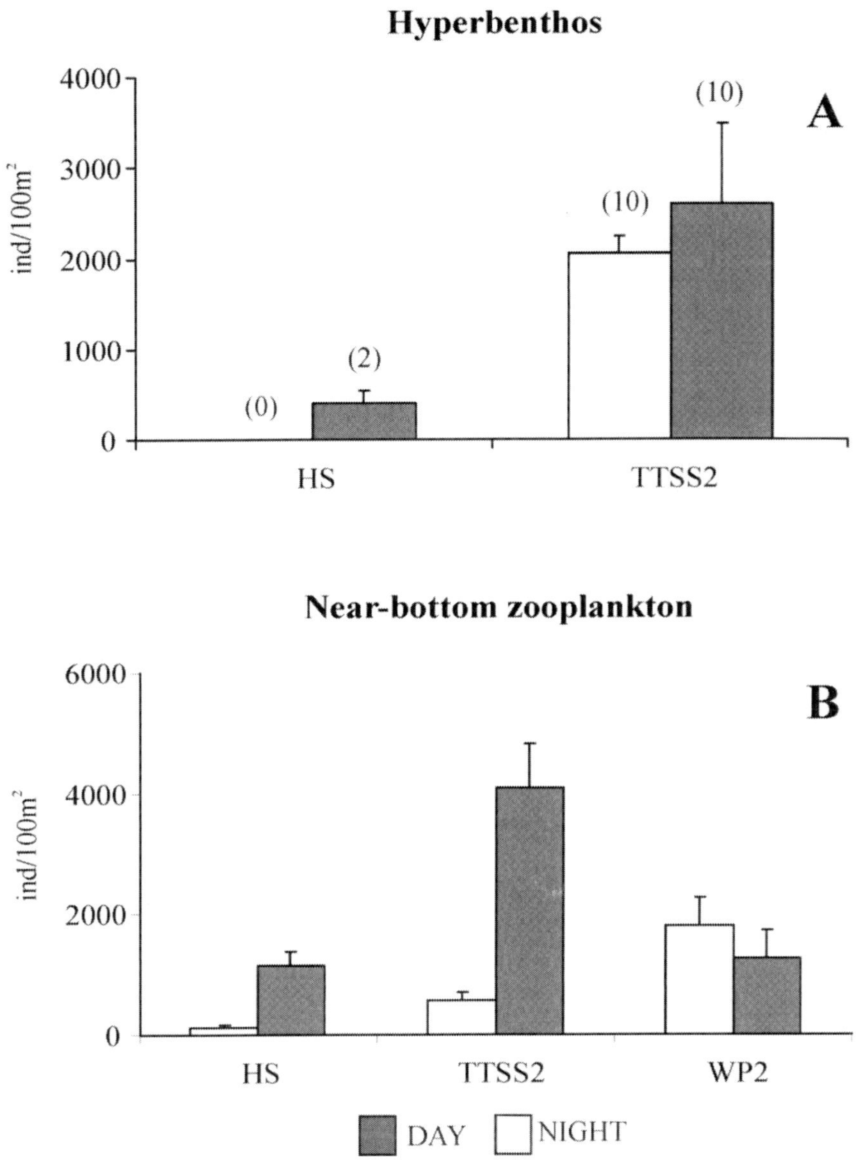

Figure 7. (A) Hyperbenthos collected with TTSS2 and HS (average±standard deviation), and (B) Near-bottom zooplankton collected with the same sledges and the WP2 plankton net (average±standard deviation), during day and night at the mid-shelf of Heraklion Bay (50 m). Expected number of species for a theoretical sample of 100 specimens, ES (100), is indicated in parentheses.

Effects of Otter Trawling on BBL Macrofauna

Bottom fishing (e.g. otter trawls) can cause long-term and large-scale perturbations to the marine ecosystem and affect seafloor biodiversity as well as relevant linkages across the sediment-water interface (Smith et al., 2000). Many studies that have adopted comparative and experimental approaches have shown that benthic infauna or large epifauna can be exposed, damaged or killed by the passage of a trawl and this may lead to increased opportunistic feeding by predators and scavengers (Jennings & Kaiser, 1998). However, there are still problems and technical difficulties in carrying out trawl impact experiments considering that the highly mobile and small sized invertebrates residing on or just above the sediment-water interface are caught only occasionally by conventional sampling gears (Eleftheriou, 2013) and are not retained in the net of traditional fisheries assessment gears (Hall, 1999).

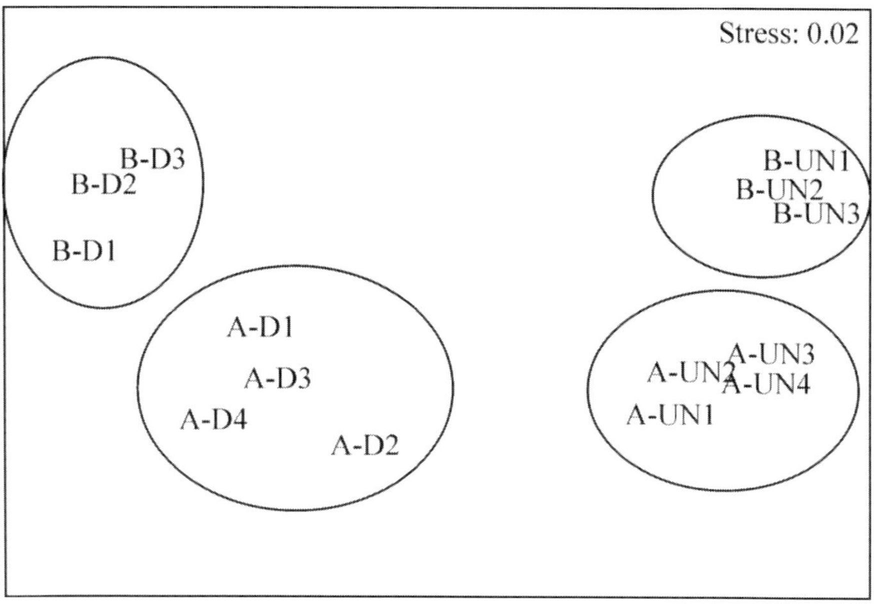

Figure 8. Multidimensional Scaling (MDS) ordination analysis based on density matrices of BBL macrofaunal taxa (R=1, $p<0.001$). B: before the beginning of the trawling season, A: after the beginning of the trawling season, D: disturbed tows of the sledge, UN: undisturbed tows of the sledge (Koulouri et al., 2005, modified).

The third set of experiments involved the application of the TTSS2 equipped with the artificial resuspension facility, which also simulates the disturbance of the seabed caused by the passage of an otter trawl groundrope, on a traditional fishing lane of Heraklion Bay (Koulouri et al., 2003; 2005). Observations made during sediment disturbance experiments showed significant perturbations of BBL macrofauna (Figure 8). Sediment disturbance using the otter trawl groundrope suspends not only a large number of epibenthic animals, but also many hyperbenthic and zooplanktonic organisms increasing their vulnerability to predators. The TTSS2 also seems to be effective in describing the available prey, exploited by local bentho-pelagic fish populations (Koulouri et al., 2012). Shifts in the diet of 9 bentho-pelagic fish species were observed due to bottom trawling disturbance and are clearly reflected in changes of BBL macrofauna (Koulouri et al., 2005; 2006; 2012). These changes include: (1) decrease of epibenthic and hyperbenthic taxa which is indicative of displacement, damage or kill of these animals by the passage of trawl gear thus generating food source for predators and scavengers, and (2) increase of several hyperbenthic and zooplanktonic taxa due to their immigration from neighbouring untrawled areas or the upper layers of the water column into the disturbed area of the seabed characterized by augmented availability of food.

CONCLUSION

Over the last decade, many facets of benthic research have been accompanied by significant technological achievements. In contrast, a slowdown in technological progress on macrofaunal techniques and a scarcity of ideas in relevant technical and methodological issues can hardly lead to the conclusion that the existing methodology concerning sampling of an extremely complex environment has reached any degree of perfection (Eleftheriou, 2013).

The new BBL sampling towed sledge system, modified to resuspend sediment surface, has proved to be suitable and effective for studying BBL macrofauna communities and the resulting ecosystem responses. The development and application of similar sampling procedures may assist in overcoming problems related to the inaccessibility of BBL environment and thus making the study of this particular habitat more comprehensive.

The upgrading of this novel sampling platform is underway. The new system will be applied for a holistic ecosystem approach of the BBL habitat aiming at investigating the ability of demersal fish species to adapt to alterations in the availability of food resources due to natural or anthropogenic disturbances in both eutrophic and oligotrophic continental shelf areas. The further application of this methodological approach is expected to give new insights towards a better understanding of the role of biodiversity related to the sediment-water interface with respect to nutrient regeneration, carbon cycling and energy transfer to higher trophic levels as well as to relevant bottom fishing gear effects.

ACKNOWLEDGMENTS

This work was carried out within the framework of the project "Development of a new method for the quantitative measurement of the effects of otter trawling on benthic nutrient fluxes and sediment biogeochemistry," which was financed by the European Commission (DG. XIV, Studies for the support of CFP). The author would like to thank Dr Costas Dounas and Prof. Anastasios Eleftheriou as well as Mrs M. Eleftheriou for critical review and comments on the manuscript.

REFERENCES

Aldred, R.G., Thurston, M.H., Rice, A.L. & Morley, D.R. (1976). An acoustically monitored opening and closing epibenthic sledge. *Deep Sea Research,* 23, 167-174.

Bergman, M.J.N. & van Santbrink, J.W. (1994). A new benthos dredge (triple-D) for quantitative sampling of infauna species of low abundance. *Netherlands Journal of Sea Research,* 33(1), 129-133.

Bergstrom, B.I. (1985). Aspects of natural foraging by Sepietta oweniana (Mollusca, Cephalopoda). *Ophelia,* 24 (1), 65-74.

Beyer, F. & Indrehus, J. (1995). *Effekter av forurensning og dypvannsutskiftning pa faunaen langs bunnen av Oslofjorden basert pa materiale samlet siden 1952.* Rapport nr. 621/95. Del II. Biologisk Institutt, Universitetet i Oslo.

Boudreau, B.P. & Jørgensen, B.B. (2001). *The benthic boundary layer: Transport processes and biogeochemistry.* Oxford University Press.

Boysen, H.O. (1975). Seasonal variations in abundance of hyperbenthic animals in the Kiel Bight. *Merentutkimuslaitoksen Julkaisul/ Havsforskningsinstitutets Skrift,* 239, 206-212.

Brandt, A. (1995). Peracarid fauna (Crustacea, Malacostraca) of the Northeast Water Polynya off Greenland: documenting close benthic-pelagic coupling in the Westwind Trough. *Marine Ecology Progress Series,* 121, 39-51.

Brandt, A. & Barthel, D. (1995). An improved supra- and epibenthic sledge for catching Peracarida (Crustacea, Malacostraca). *Ophelia,* 43(1), 15-23.

Brandt, A. & Schnack, K. (1999). Macrofaunal abundance at 79ºN off East Greenland: opposing data from epibenthic-sledge and box-corer samples. *Polar Biology,* 22, 75-81.

Brandt, A. & Berge, J. (2007). Peracarid composition, diversity and species richness in the area of the Northeast Water polynya, East Greenland (Crustacea, Malacostraca). *Polar Biology,* 31, 15-22.

Brattegard, T. & Fossa, J.H. (1991). Replicability of an epibenthic sampler. *Journal of Marine Biological association of the United Kingdom,* 71, 153-166.

Brunel, P., Besner, M., Messier, D., Poirier, L., Granger, D. & Weinstein, M. (1978). Le traîneau Macer-GIROQ: appareil amélioré pour l'échantillonnage quantitatif de la petite faune nageuse au voisinage du fond. *Internationale Revue* der gesamten *Hydrobiologie* und Hydrographie, 63(6), 815-829.

Buhl-Jensen, L. (1986). The benthic amphipod fauna of the West-Norwegian continental shelf compared with the fauna of five adjacent fjords. *Sarsia,* 71, 193-208.

Buhl-Jensen, L. & Fossa, J.H. (1991). Hyperbenthic crustacean fauna of the Gullmarfjord area (western Sweden): species richness, seasonal variation and long-term changes. *Marine Biology,* 109, 245-258.

Carleton, J.H. & Hamner, W.M. (1989). Resident mysids: community structure, abundance and small-scale distributions in a coral reef lagoon. *Marine Biology,* 102, 461-472.

Cartes, J.E. (1993a). Feeding habits of pasiphaeid shrimps close to the bottom on the Western Mediterranean slope. *Marine Biology,* 117, 459-468.

Cartes, J.E. (1993b). Feeding habits of oplophorid shrimps in the deep western Mediterranean. *Journal of the Marine Biological Association of the United Kingdom,* 73, 193-206.

Cartes, J.E. (1993c). Diets of deep-sea brachyuran crabs in the Western Mediterranean Sea. *Marine Biology,* 117, 449-457.

Cartes, J.E., Company, J.B. & Maynou, F. (1994a). Deep-water decapod crustacean communities in the Northwestern Mediterranean: influence of submarine canyons and season. *Marine Biology,* 120, 221-229.

Cartes, J.E., Sorbe, J.C. & Sarda, F. (1994b). Spatial distribution and swimming activity of deep-sea decapods and euphausiids near the bottom in the northwestern Mediterranean. *Journal of Experimental Marine Biology and Ecology, 179*, 131-144.

Cartes, J.E. & Maynou, F. (1998). Food consumption by bathyal decapod crustacean assemblages in the Western Mediterranean: predatory impact of megafauna and the food consumption-food supply balance in a deep-water food web. *Marine Ecology Progress Series*, 171, 233-246.

Cartes, J.E., Abelló, P., Lloris, D., Carbonell, A., Torres, P., Maynou, F. & de Sola, L.G. (2002). Feeding guilds of western Mediterranean demersal fish and crustaceans: an analysis based on a spring survey. *Scientia Marina, 66 (Suppl.* 2), 209-220.

Cartes, J.E., Papiol, V., Palanques, A., Guillén, J. & Demestre, M. (2007). Dynamics of suprabenthos off the Ebro Delta (Catalan Sea: western Mediterranean): Spatial and temporal patterns and relationships with environmental factors. *Estuarine, Coastal and Shelf Science,* 75, 501-515.

Champalbert, G. & Macquart-Moulin, C. (1970). *Les peracarides de l' hyponeuston nocturne du Golfe de Marseille. Cahiers de Biologie* Marine, 11, 1-29.

Chevrier, A., Brunel P. & Wildish, D.J. (1991). Structure of a suprabenthic shelf sub-community of gammaridean amphipoda in the Bay of Fundy compared with similar sub-communities in the Gulf of St. Lawrence. *Hydrobiologia,* 223, 81-104.

Christiansen, B. & Nuppenau, V. (1997). The IHF Fototrawl: experiences with a television-controlled, deep sea epibenthic sledge. *Deep Sea Research I,* 44(3), 533-540.

Cornet, M., Lissalde, J.P., Bouchet, J.M., Sorbe, J.C. & Amoureux, L. (1983). Données qualitatives sur le benthos et le suprabenthos d' un transect du plateau continental sud-Gascogne. *Cahiers de Biologie Marine,* 24, 69-84.

Cunha, M.R., Sorbe, J.C. & Bernardes, C. (1997a). On the structure of the neritic suprabenthic communities from the Portuguese continental margin. *Marine Ecology Progress Series,* 157, 119-137.

Cunha, M.R., Sorbe, J.C. & Bernardes, C. (1997b). On the structure of the coastal suprabenthic communities from the continental shelf off Aveiro (NW Portugal). *Arquivos do Museu Bocage Nova Série,* 3 (3), 165-188.

Dauvin, J.C. & Lorgeré, J.C. (1989). Modifications du traîneau Macer-GIROQ pour l' amélioration de l' échantillonage quantitatif étagé de la faune suprabenthique. *Journal de la Recherche Océanographique,* 14, 65-67.

Dauvin, J., Iglesias, A. & Lorgeré, J. (1994). Circalittoral suprabenthic coarse sand community from the Western English Channel. *Journal of Marine Biological Association of the United Kingdom,* 74, 543-562.

Dauvin, J., Sorbe, J. & Lorgeré, J. (1995). Benthic boundary layer macrofauna from the upper continental slope and the Cap Ferret canyon (Bay of Biscay). *Oceanologica Acta,* 18(1), 113-122.

Dauvin, J.C. & Zouhiri, S. (1996). Suprabenthic crustacean fauna of a dense Ampelisca community from the English channel. *Journal of the Marine Biological Association of the United Kingdom,* 76, 909-929.

Dauvin, J.C. & Vallet, C. (2006). The near-bottom layer as an ecological boundary in marine ecosystems: diversity, taxonomic composition and community definitions. *Hydrobiologia,* 555, 49-58.

Dauvin, J.C., Vallet, C., Mouny, P. & Zouhiri, S. 2000. Main characteristics of the boundary layer macrofauna in the English Channel. *Hydrobiologia,* 426, 139-156.

Dewicke, A., Cattrijsse, A., Mees, J. & Vincx, M. (2002). Evidence for an enriched hyperbenthic fauna in the Frisian front (North Sea*). Journal of Sea Research,* 47, 121-139.

Dewicke, A., Cattrijsse, A., Mees, J. & Vincx, M. (2003). Spatial patterns of the hyperbenthos of subtidal sandbanks in the southern North Sea. *Journal of Sea Research,* 49, 27-45.

Dounas, C. (2006). A new apparatus for the direct measurement of the effects of otter trawling on benthic nutrient releases. *Journal of Experimental Marine Biology and Ecology,* 339(2), 251-259.

Dounas, C.G., Davies, I.M., Hayes, P.J., Arvanitidis, C.D. & Koulouri, P.T. (2002). *Development of a new method for the quantitative measurement of the effect of otter trawling on benthic nutrient fluxes and sediment biogeochemistry.* European Commission, Study Project Number 99/036, Final Technical Report, HCMR, Heraklion, Crete, Greece.

Dounas, C., Davies, I., Hayes, P., Arvanitidis, C. & Koulouri, P. (2005). The effect of different types of otter trawl groundrope on benthic nutrient fluxes and sediment biogeochemistry. American Fisheries Society Symposium, 41, 539-544.

Dounas, C., Davies, I., Triantafyllou, G., Koulouri, P., Petihakis, G., Arvanitidis, C., Sourlatzis, G. & Eleftheriou, A. (2007). Large scale impacts of bottom trawling on shelf primary productivity. *Continental Shelf Research,* 27, 2198-2210.

Eleftheriou, A. (2013). *Methods for the study of marine benthos.* 4th Edition. Oxford, UK, John Wiley & Sons, Ltd.

Fanelli, E., Cartes, J.E., Badalamenti, F., Rumolo, P. & Sprovieri, M. (2009a). Trophodynamics of suprabenthic fauna on coastal muddy bottoms of the southern Tyrrhenian Sea (western Mediterranean). *Journal of Sea Research,* 61, 174-187.

Fanelli, E., Cartes, J.E., Rumolo, P. & Sprovieri, M. (2009b). Food-web structure and trophodynamics of mesopelagic-suprabenthic bathyal macrofauna of the Algerian Basin based on stable isotopes of carbon and nitrogen. *Deep Sea Research I,* 56, 1504-1520.

Fanelli, E., Cartes, J.E., Badalamenti, F., D'Anna, G., Pipitone, C., Azzurro, E., Rumolo, P. & Sprovieri, M. (2011). Meso-scale variability of coastal suprabenthic communities in the southern Tyrrhenian Sea (western Mediterranean). *Estuarine, Coastal and Shelf Science,* 91, 351-360.

Fossa, J.H. (1985). Near-bottom vertical zonation during daytime of deep-living hyperbenthic mysids (Crustacea: Mysidacea). *Sarsia,* 70, 297-307.

Fossa, J.H., Larsson, J. & Buhl-Jensen, L. (1988). A pneumatic, bottom activated, opening and closing device for epibenthic sledges. *Sarsia,* 73, 299-302.

Hall, S.J. (1999). *The Effects of Fishing on marine ecosystems and communities.* Oxford, UK, Blackwell Science Ltd.

Hessler, R.R. & Sanders, H.L. (1967). Faunal diversity in the deep-sea. *Deep Sea Research* 14, 65-78.

Hesthagen, I.H. (1973). Diurnal and seasonal variations in the near-bottom fauna-the hyperbenthos-in one of the deeper channels of the Kieler Bucht (Western Baltic). *Kieler Meeresforschungen,* 29, 116-140.

Hesthagen, I.H. & Gjermundsen, B. (1978). The replicability of sampling the hyperbenthic region by means of Beyer's 50 cm epibenthic closing net. *Kieler Meeresforschungen,* 26, 1-10.

Hesthagen, I.H. & Gjermundsen, B. (1979). Late summer diurnal migration in the hyperbenthos of Vejsnas Rinne, Western Baltic. *Kieler Meeresforschungen,* 27, 19-26.

Huberdeau, L. & Brunel, P. (1982). Efficiency and comparative faunistic selectivity of four endo-, epi- and suprabenthic samplers on two bottom types. *Marine Biology* 69, 331-343.

Jennings, S. & Kaiser, M.J. (1998). The effects of fishing on marine ecosystems. *Advances in Marine Biology,* 34, 201-352.

Kaartvedt, S. (1985). Diel changes in small-scale vertical distribution of hyperbenthic mysids. *Sarsia,* 70, 287-295.

Kaiser, M.J., Rogers, S.I. & McCandless, D.T. (1994). Improving quantitative surveys of epibenthic communities using a modified 2 m beam trawl. *Marine Ecology Progress Series,* 106, 131-138.

Karakassis, I. & Eleftheriou, A. (1997). The continental shelf of Crete: structure of macrobenthic communities. *Marine Ecology progress Series,* 160, 185-196.

Kim, S.L. & Oliver, J.S. (1989). Swarming benthic crustaceans in the Bering and Chukchi seas and their relation to geographic patterns in gray whale feeding. *Canadian Journal of Zoology,* 67, 1531-1542.

Koulouri, P., Dounas, C. & Eleftheriou, A. (2003). A new apparatus for the direct measurement of otter trawling effects on the epibenthic and hyperbenthic macrofauna. *Journal of the Marine Biological Association U.K.,* 83, 1363-1368.

Koulouri, P., Dounas, C. & Eleftheriou, A. (2005). Preliminary results on the effect of otter trawling on hyperbenthic communities in Heraklion Bay (Eastern Mediterranean, Cretan Sea). *American Fisheries Society Symposium,* 41, 529-537.

Koulouri, P., Dounas, C. & Eleftheriou, A. (2006). *New methodologies for the study of Bottom Trawling Effects on Deep-Sea BBL habitats.* 2006 Ocean Sciences, 20-24 February 2006, Honolulu, Hawaii, U.S.A.

Koulouri, P., Dounas, C., Radin, F. & Eleftheriou, A. (2009). Near-bottom zooplankton in the continental shelf and upper slope of Heraklion Bay (Crete, Greece, Eastern Mediterranean): observations on vertical distribution patterns. *Journal of Plankton Research,* 31 (7), 753-762.

Koulouri, P., Arvanitidis, C., Koutsoubas, D., Dounas, C. & Eleftheriou, A. (2011). New field methodological approaches for the study of benthic boundary layer (BBL) biodiversity. *World Conference on Marine Biodiversity,* 26-30 September 2011, Aberdeen, Scotland, U.K.

Koulouri, P., Dounas, C., Arvanitidis, C., Koutsoubas, D. & Eleftheriou, A. (2012). Effects of otter trawling on benthic boundary layer (BBL) macrofauna and implications for the diet of bentho-pelagic fish species in a fishing ground of eastern Mediterranean. *6th World Fisheries Congress,* 7-11 May 2012, Edinburgh, Scotland, U.K.

Koulouri, P., Dounas, C. & Eleftheriou, A. (2013). Hyperbenthic community structure over oligotrophic continental shelves and slopes: Crete (South Aegean Sea, NE Mediterranean). *Estuarine, Coastal and Shelf Science,* 117, 188-198.

Labropoulou, M. & Eleftheriou, A. (1997). The foraging ecology of two pairs of congeneric demersal fish species: importance of morphological characteristics in prey selection. *Journal of Fish Biology,* 50, 324-340.

Labropoulou, M. & Kostikas, I. (1999). Patterns of resource use in deep-water decapods. *Marine Ecology Progress Series,* 184, 171-182.

Ligas, A., Sartor, P., Sartini, M., Vanucci, A., De Ranieri, S. & Cartes, J.E. (2007). Hyperbenthic Mysidae Haworth, 1825 (Peracarida, Mysida) from the continental shelf of the northern Adriatic Sea. *Crustaceana,* 80 (8), 969-977.

Macer, C.T. (1967). A new bottom-plankton sampler. *Journal du Conseil International pour l' Exploration de la Mer,* 31(2), 158-163.

Marshall, N.B. & Merrett, N.R. *The existence of a benthopelagic fauna in the deep sea.* In: Angel M., editor, A Voyage of Discovery: George Deacon 70th Anniversary, Oxford, UK, Pergamon Press Ltd, 1977, 483-497.

Martin, B. & Christiansen, B. (1997). Diets and standing stocks of benthopelagic fishes at two bathymetrically different mid-oceanic localities in the northeast Atlantic. *Deep-Sea Research I,* 44(4), 541-558.

Mees, J. & Jones, M.B. (1997). The Hyperbenthos. *Oceanography and Marine Biology: An Annual Review,* 35, 221-255.

Ockelmann, K.W. (1964). An improved detritus-sledge for collecting meiobenthos. *Ophelia,* 1(2), 217-222.

Oug, E. (1977). Faunal distribution close to the sediment of a shallow marine environment. *Sarsia,* 63, 115-121.

Psarra, S., Tselepides, A. & Ignatiades, L. (2000). Primary productivity in the oligotrophic Cretan Sea (NE Mediterranean): seasonal and interannual variability. *Progress in Oceanography,* 46, 187-204.

Rice, A.L., Aldred, R.G., Darlington, E. & Wild, R.A. (1982). The quantitative estimation of the deep-sea megabenthos; a new approach to an old problem. *Oceanologica Acta,* 5(1), 63-72.

Rothlisberg, P.C. & Pearcy, W.G. (1977). An epibenthic sampler used to study the ontogeny of vertical migration of Pandalus jordani (Decapoda, Caridea). *Fisheries Bulletin of the United States,* 74, 994-997.

Rowell, T.W., Schwinghamer, P., Chin-Yee, M., Gilkinson, K., Gordon Jr, D.C., Hartgers, E., Hawryluk, M., McKeown, D.L., Prena, J., Reimer, D.P., Sonnichsen, G., Steeves, G., Vass, W.P., Vine, R. & Woo, P. (1997). Grand Banks otter trawling experiment: III. Sampling equipment, experimental design, and methodology. *Canadian Technical Report of Fisheries and Aquatic Sciences 2190*, viii+36p.

Sainte-Marie, B. & Brunel, P. (1983). Differences in life history and success between suprabenthic shelf populations of Arrhis phyllonyx (Amphipoda Gammaridea) in two ecosystems of the Gulf of St. Lawrence. *Journal of Crustacean Biology,* 3(1), 45-69.

Sainte-Marie, B. & Brunel, P. (1985). Suprabenthic gradients of swimming activity by cold-water gammaridean amphipod Crustacea over a muddy shelf in the Gulf of Saint Lawrence. *Marine Ecology Progress Series,* 23, 57-69.

San Vicente, C., Ramos, A., Jimeno, A. & Sorbe, J.C. (1997). Suprabenthic assemblages from South Shetland islands and Bransfield Strait (Antarctica): preliminary observations on faunistical composition, bathymetric and near-bottom distribution. *Polar Biology,* 18, 415-422.

San Vicente, C., Ramos, A. & Sorbe, J.C. (2006). Suprabenthic euphausiids and mysids from the South Shetland islands and the Bransfield Strait, Southern ocean (BENTERT-95 cruise). *Polar Biology,* 29, 211-222.

San Vicente, C., Munilla, T., Corbera, J., Sorbe, J.C. & Ramos, A. (2009). Suprabenthic fauna from the Bellingshausen Sea and western Antarctic Peninsula: spatial distribution and community structure. *Scientia Marina* 73 (2), 357-368.

Shand, C.W. & Priestley, R. (1999). *A towed sledge for benthic surveys. Fisheries Research Services,* Scottish Fisheries Information Pamphlet, no. 22, 8 pp.

Smith, C.R., Austen, M.C., Boucher, G., Heip, C., Hutchings, A., King, G.M., Koike, I., Lambshead, P.J.D. & Snelgrove, P. (2000). Global change and biodiversity linkages across the sediment-water interface. *Bioscience,* 50(12), 1108-1120.

Sorbe, J.C. (1981). Rôle du benthos dans le régime alimentaire des poissons démer-saux du secteur Sud Gascogne (Role of benthos in the diet of demersal fishes from South Gascogne area). *Kieler Meeresforschungen sonderheft,* 5, 479-489.

Sorbe, J.C. (1983). Description d' un traîneau destiné à l' échantillonnage quantitatif étagé de la faune suprabenthique néritique. *Annales de l' Institut Océanographique Paris,* 59 (2), 117-126.

Sorbe, J.C. (1989). Structural evolution of two suprabenthic soft-bottom communities of the South Gascogne continental shelf. *Scientia Marina,* 53(2-3), 335-342.

Thouzeau, G. & Vine, R.D. (1991). L' échantillonnage du mégabenthos en zone hauturière: technique développée sur le Georges Bank (Atlantique

Nord-ouest). *Comptes Rendus de l' Académie des Sciences Paris,* 312 (SERIE III), 607-613.

Tselepides, A., Zervakis, V., Polychronaki, T., Danovaro, R. & Chronis, G. (2000). Distribution of nutrients and particulate organic matter in relation to the prevailing hydrographic features of the Cretan Sea (NE Mediterranean). *Progress in Oceanography,* 46, 113-142.

Turnbull, C.T. & Watson, R.A. (1992). Sledges for daytime sampling of juvenile penaeid shrimp. *Fisheries Research,* 14, 31-40.

Vallet, C. & Dauvin, J. (1998). Composition and diversity of the benthic boundary layer macrofauna from the English Channel. *Journal of Marine Biological Association of the United Kingdom,* 78(2), 387-409.

Vallet, C. & Dauvin, J.C. (1999). Seasonal changes of macrozooplankton and benthic boundary layer macrofauna from the Bay of Saint-Brieuc (Western English Channel). *Journal of Plankton Research,* 21(1), 35-49.

Villanueva, R. & Guerra, A. (1991). Food and prey detection in two deep-sea cephalopods: Opisthoteuthis agassizi and O. vossi (Octopoda:Cirrata). *Bulletin of Marine Science,* 49(1-2), 288-299.

Wang, Z. & Dauvin, J.C. (1994). The suprabenthic crustacean fauna of the infralittoral fine sand community in the Bay of Seine (Eastern English Channel): composition, swimming activity and diurnal variation. *Cahiers de Biologie Marine,* 35, 135-155.

Wildish, D.J., Wilson, A.J. & Frost, B. (1992). Benthic boundary layer macrofauna of Browns Bank, Northwest Atlantic, as potential prey of juvenile benthic fish. *Canadian Journal of Fisheries and Aquatic Sciences,* 49, 91-98.

Wishner, K. F. (1980). The biomass of the deep-sea benthopelagic plankton. *Deep-Sea Research,* 27A, 203-216.

Youngbluth, M.J. (1982). Sampling demersal zooplankton: a comparison of field collections using three different emergence traps. *Journal of Experimental Marine Biology and Ecology,* 61, 111-124.

Zouhiri, S. & Dauvin, J.C. (1996). Diel changes of the benthic boundary layer macrofauna over coarse sand sediment in the western English Channel. *Oceanologica Acta,* 19(2), 141-153.

INDEX

A

access, 137
accommodation, viii, 2, 4
age, 6, 9, 10, 11, 13, 15, 22, 37, 42, 43, 46, 50, 55, 61, 79, 80, 88, 89, 90, 95, 96, 101, 102, 106, 110, 151, 152, 154, 157
aggregation, 136
algae, 87, 160
amplitude, 38, 43, 52, 128, 165, 177
apex, 79
assessment, 110, 117, 180
asymmetry, 86, 94
attribution, 55, 69, 82

B

banks, 7, 10, 23, 25, 27, 49, 50
base, 3, 17, 18, 44, 80, 88, 89, 99, 113, 125, 140, 147
beams, 14
benthic boundary layer (BBL), ix, 164, 187
biodiversity, x, 164, 180, 182, 187, 189
biomass, 165, 190
bounds, 48
breakdown, 90, 102

C

Cairo, 123, 158, 161, 162
calibration, 6
canyon formation, viii, 2, 17, 18
capillary, 128
carbon, x, 164, 182, 186
case study(ies), 107, 120, 121, 162
cell size, 155
CFP, 182
chemical, 108
Chile, 27, 160
chlorophyll, 172
circulation, 83, 147
classes, 37
classification, 113
climate, 143, 144, 147, 156
commercial, x, 164, 166
community(ies), vii, x, 90, 136, 138, 164, 166, 169, 171, 173, 174, 181, 183, 184, 185, 186, 187, 189, 190
compaction, 151
composition, 8, 9, 16, 83, 90, 174, 183, 185, 189, 190
comprehension, 19
compression, 140
condensation, 83
conditioning, viii, 2, 49
configuration, 82, 84, 138, 149, 156

Congress, 103, 109, 187
consolidation, 50
construction, 3, 94
consumption, 184
contamination, 166
content analysis, 174
continental, i, iii, v, vii, 1, 108, 116, 136, 163, 171, 172, 185
contour, 19, 21, 125, 129
coral reefs, ix, 123, 127, 133, 137, 155, 159
correlation, 13, 18, 40, 44, 48, 50, 55, 85, 154
correlations, 52
corrosion, 86
covering, 3, 52, 94, 136
crabs, 183
cracks, 28
creep, viii, 2, 12, 14, 29, 30, 38
crises, 143
crop(s), 7, 8, 10, 11, 58, 61, 62, 66, 69, 96, 99, 101
crust, 7
cycles, 8, 49, 62, 86, 118
cycling, x, 164, 182

D

data analysis, 49, 96, 128
data set, viii, 123, 128, 131, 155
database, 129
database management, 129
defence, 88
deformation, 6, 15, 102, 111
Delta, 171, 184
density matrices, 180
deposition, 18, 23, 44, 92, 118, 120, 136, 138, 143
depression, ix, 9, 27, 28, 98, 101, 124, 138, 143, 156
depth, viii, 2, 3, 14, 15, 17, 21, 22, 25, 37, 38, 43, 46, 61, 62, 69, 80, 81, 83, 84, 94, 95, 101, 124, 126, 128, 133, 140, 141, 147, 151, 156, 162
destructive process, 133
detachment, 33, 39
detection, viii, 123, 155, 190
deviation, 179
diet, 181, 187, 189
diffusion, 83
direct measure, 185, 187
direct observation, 101
discharges, 172
displacement, 71, 181
distribution, vii, viii, 2, 4, 23, 82, 83, 98, 125, 133, 165, 174, 177, 178, 184, 186, 187, 188, 189
diversity, x, 164, 165, 174, 175, 176, 183, 185, 186, 190
drainage, ix, 6, 13, 14, 22, 25, 30, 48, 83, 124, 126, 142, 144, 146, 147, 149, 150, 151, 156, 161

E

earthquakes, 23, 28, 36, 115
ecology, 188
ecosystem, ix, 163, 180, 181
Egypt, 123, 125, 128, 142, 145, 147, 150, 157, 158, 159, 160, 161, 162
electric current, 166
energy, ix, 28, 35, 36, 52, 133, 163, 182
energy transfer, x, 164, 182
England, 125
environment(s), viii, 2, 4, 36, 55, 78, 80, 81, 82, 83, 84, 88, 89, 90, 102, 128, 133, 159, 161, 166, 173, 174, 177, 181
environmental conditions, 83, 172, 173
environmental factors, 184
epibenthic communities, 187
equilibrium, 83
equipment, 188
erosion, 6, 15, 18, 22, 23, 24, 25, 29, 37, 38, 79, 80, 86, 88, 92, 120, 126, 137, 143, 147, 149
EU, x, 164, 169
Eurasia, 112
Europe, 112
European Commission, 182, 185
evacuation, 114
evaporation, 149

evidence, 7, 15, 17, 22, 27, 32, 37, 38, 39, 52, 71, 81, 85, 102, 107, 108, 118, 119, 120, 147
evolution, viii, ix, 2, 4, 11, 18, 23, 25, 33, 38, 39, 50, 53, 72, 84, 85, 103, 108, 112, 114, 115, 117, 124, 128, 149, 157, 158, 159, 160, 189
experimental design, 188
exposure, 15, 37
extraction, 161, 162

F

facies, 5, 11, 28, 30, 31, 32, 33, 38, 53, 54, 69, 70, 71, 73, 74, 75, 78, 83, 92, 101, 113
fauna, ix, 163, 165, 171, 173, 174, 177, 183, 185, 186, 188, 189, 190
fish, x, 164, 165, 166, 174, 181, 184, 187, 188, 190
fisheries, 166, 180
fishing, 180, 181, 182, 186, 187
flank, 7, 19, 24, 26, 27, 28, 29, 32, 46, 48, 49, 69, 96, 107
flooding, 18, 38, 53, 81, 133, 160
flora, 160
flora and fauna, 160
fluctuations, 38
fluid, 19, 109
food(s), x, 145, 151, 164, 165, 176, 177, 181, 184
food web, x, 164, 165, 184
force, 177
formation, viii, 2, 6, 8, 17, 18, 19, 50, 79, 82, 86, 106, 135, 140, 157
fractures, 28, 140
fragments, 8, 83, 89, 90, 102
freshwater, 172

G

gemma, 1
Geographic Information System (GIS), 36, 158, 160
geology, 3, 6, 15, 52, 103, 115, 159, 161
geometry, 28, 33, 53, 70, 73, 74, 75
Germany, 123, 143
glasses, 119
GPS, 168
grades, 46, 55, 62, 91
grading, 17, 100
grasses, 85
gravity, 4, 13, 14, 17, 18, 19, 26, 29, 31, 92, 114, 125, 128, 138, 145, 157
Great Britain, 160
Greece, 163, 172, 185, 187
grids, 17
grouping, 4
growth, 22, 98, 133
GSA, 114, 115, 121
guidelines, vii, 1, 4
Guinea, 120
Gulf of Mexico, 113

H

habitat(s), ix, 163, 165, 170, 174, 181, 187
Hawaii, 27, 187
height, 7, 11, 37, 128, 167, 168
highlands, 145
history, 36, 100, 102, 103, 106, 116, 117, 162, 189
Holocene, viii, 2, 7, 10, 11, 13, 14, 15, 24, 31, 37, 38, 39, 42, 43, 46, 53, 61, 62, 73, 83, 88, 89, 90, 92, 99, 101, 102, 106, 138, 151, 156, 159
human, 161
hydrothermal activity, 9
hypothesis, 23, 33, 121

I

ideal, 156
identification, 15, 27, 37, 50, 53, 54, 96, 129
illumination, 177
image(s), 3, 33, 84, 108, 128, 129, 130, 139, 155
image interpretation, 139

imagery, vii, 1, 27, 32
immigration, 181
impulsive, 39
individuals, 166, 177
inferences, vii, 3, 104
initiation, 18, 118
integration, viii, 5, 17, 123
interface, ix, 163, 164, 177, 180, 182, 189
interference, 15, 37
intrusions, 53, 138
inversion, 83
invertebrates, 180
islands, 4, 6, 7, 16, 22, 27, 31, 32, 35, 36, 49, 53, 69, 96, 129, 135, 140, 142, 159, 189
isolation, ix, 163
isotope, 108
Israel, 161
issues, 181
Italy, v, vii, 1, 3, 13, 16, 36, 103, 104, 105, 106, 107, 108, 109, 110, 111, 112, 114, 115, 116, 117, 118, 119, 120

J

Japan, 27
Jordan, 161

K

kill, 181

L

larvae, 165, 174, 175, 176, 177
Late Pleistocene, viii, 2, 4, 6, 12, 14, 18, 31, 43, 45, 46, 50, 61, 86, 88, 90, 91, 92, 96, 102, 106
lead, 180, 181
levees, 21, 31, 55, 73
light, ix, 102, 112, 124, 137, 157, 172
limestone, 89, 119
lithological units, 89, 90
lithology, 84
Luo, 157

M

magnetic field, 44, 97
magnetization, 50
magnitude, 175
management, 162
mantle, 88, 140
mapping, vii, 1, 3, 5, 14, 82, 88, 100, 110, 162
Marani, 112
marine benthos, 185
marine ecosystem, ix, 163, 166, 180, 185, 186
marine environment, 188
Mars, 140, 160
mass, 24, 28, 33, 34, 39, 58, 81, 116, 119, 121
materials, ix, 24, 39, 58, 98, 127, 163
matrix, 28, 32, 36, 83, 88, 89, 90, 102, 119
matter, ix, 164, 165, 173, 177
measurement, 182, 185
median, 125
Mediterranean, 36, 103, 108, 109, 112, 143, 156, 171, 172, 173, 176, 183, 184, 186, 187, 188, 190
melt, 152
Messinian event, ix, 124, 156
meter, 90, 129
methodology, 53, 181, 188
migration, viii, 2, 4, 18, 156, 186, 188
military, viii, 123, 129
Miocene, 79, 85, 102, 125, 138, 140, 143, 144, 147, 156
mixing, 172
modelling, 33, 106
models, 18, 28, 110, 155
modifications, 149, 171
morphology, 17, 18, 27, 33, 37, 49, 50, 79, 82, 83, 84, 101, 114, 119, 136, 142, 150
morphometric, 24, 38, 84
multiple factors, 149

N

National Research Council, 13, 151
Nd, 108
Netherlands, 160, 182
New England, 116
Nile, 143
nitrogen, 186
NOAA, 157
nodes, 35
Norway, 107, 114, 121
nutrient(s), x, 84, 164, 169, 182, 185, 190

O

oceans, 18
operations, 125
opportunities, 121
organic matter, 18, 38, 177, 190
overlap, 14
overlying paleosol, 11

P

parallel, 22, 27, 43, 44, 46, 53, 66, 69, 71, 73, 74, 75, 76, 77, 81, 84, 91, 95, 99, 100, 125, 133, 137, 138, 139, 140
peat, 13
permit, 105
personal communication, 143
petroleum, 112, 113, 157, 162
Phlegrean Fields, viii, 2, 6, 7, 9, 10, 11, 12, 13, 17, 22, 39, 44, 45, 47, 49, 50, 96, 104, 116, 118, 119
photomosaics, 3, 32
physical and mechanical properties, 13
pigmentation, 83
planets, 109
plankton, x, 164, 165, 166, 169, 173, 179, 188, 190
platform, 18, 27, 30, 69, 78, 86, 87, 101, 133, 134, 136, 137, 181
playing, 18
Pliocene, 7, 119, 138
pools, 148, 149
Portugal, 184
predators, 166, 177, 180, 181
principles, 4
project, x, 107, 164, 169, 182
propagation, 32, 33, 34, 35, 49

Q

quality control, 129
quantitative estimation, 188

R

radar, 129
radio, 125
recognition, vii, 2, 4, 53
reconstruction, vii, 2, 3, 6, 13, 31, 33, 80, 94, 104, 106, 110
regeneration, x, 164, 182
regression, 50, 70, 72, 74, 75, 155
regression analysis, 155
relief, 17, 31, 33, 73, 84, 98, 101, 126, 143, 151
remote sensing, 155, 160, 162
researchers, 171
residual error, 129
resolution, vii, 1, 3, 13, 15, 16, 19, 37, 43, 51, 52, 70, 95, 96, 103, 112, 129, 130, 155
resources, x, 164, 165, 176, 181
response, 115, 133, 177
rings, 9
risk, 36
river flows, 19
river systems, 18
root, 130
root-mean-square, 130
Russia, 106

S

salinity, 125, 133, 143, 147, 149, 160, 161
salts, 147

saturation, 28
Saudi Arabia, 126, 158
savannah, 147, 156
scarcity, 149, 181
scavengers, 180, 181
science, 115
sea bottom samples, vii, 1, 3, 78, 84
sea-level rise, ix, 11, 124
seasonal changes, 128, 177
sedimentation, viii, 2, 7, 18, 19, 23, 38, 39, 53, 83, 97, 112, 145
seismic data, 34, 125
selectivity, 186
seminars, 162
shape, 21, 23, 25, 31, 33, 43, 46, 47, 49, 50, 73, 87, 133, 139, 155, 156
shelter, 177
shoreline, 44, 79, 80, 85, 86, 87, 88, 99, 125, 161
showing, viii, ix, 2, 5, 15, 18, 19, 24, 25, 27, 33, 38, 39, 40, 46, 50, 66, 74, 84, 85, 91, 93, 94, 95, 101, 124, 127, 132, 137, 141, 150, 152, 156
shrimp, 190
Side Scan Sonar imagery, vii, 1
signals, 121
simulation(s), 29, 32, 34, 38, 113, 120, 121
Sinai, 154, 159, 161
Singapore, 158
single-channel seismic profiles, vii, 2
South Africa, 120
species, ix, 164, 165, 174, 175, 176, 177, 178, 179, 181, 182, 183, 187, 188
species richness, 183
stability, 109
standard deviation, 179
state, 33, 115
stimulation, 166
stomach, 174
stratification, 53, 81
stretching, 62
strong volcanic activity, 12
structural characteristics, 28, 85
structure, vii, ix, 10, 15, 17, 22, 28, 39, 41, 44, 45, 54, 55, 66, 81, 82, 86, 92, 95, 100, 101, 102, 104, 124, 135, 136, 140, 163, 171, 173, 175, 183, 184, 186, 187, 189
style, 117
succession, viii, 2, 4, 11, 13, 31, 62, 78, 89
Sudan, 126
superimposition, 79
susceptibility, 110
Sweden, 183
symmetry, 140

T

Task Force, 103
taxa, x, 164, 174, 177, 180, 181
techniques, ix, 3, 4, 61, 155, 163, 181
technological progress, 181
technologies, 104
temperature, 133
tension, 140
terraces, viii, ix, 2, 11, 24, 38, 80, 82, 85, 86, 101, 124, 126, 151, 152, 153, 154, 155, 156, 158
territory, 37
texture, 85, 89, 90
thinning, 38, 79
three-dimensional reconstruction, 15
trace elements, 8
transducer, 14
transport, 18, 22, 29, 116, 126, 151
transportation, 126
tuff, 9, 11, 13, 14, 29, 31, 62
Turkey, 140

U

U.S. Geological Survey, 119, 121
UN, 180
United Kingdom (UK), 124, 145, 168, 183, 185, 186, 188, 190
United States, 114, 188
uranium, 158, 159
urban, 172
urbanization, 39

V

variations, viii, 2, 4, 6, 8, 52, 86, 94, 149, 182, 186
vector, 130
vegetation, 151
velocity, 34, 128
volcanism, viii, 2, 6, 10, 15, 17, 37, 39, 46, 106, 109, 115
vulnerability, 181

W

Washington, 27, 112, 114, 117, 119, 121, 160
waste, 172
watershed, 162
web, x, 164, 186
wells, 6, 129
wind speed, 128
withdrawal, 117
wood, 58

Y

Yemen, 143

Z

zooplankton, 165, 174, 176, 177, 179, 187, 190